Product Design and the Role of Representation

Product Design and the Role of Representation

Foundations for Design Thinking in Practice

Eujin Pei and James Andrew Self

CRC Press
Taylor & Francis Group
Boca Raton London New York

CRC Press is an imprint of the
Taylor & Francis Group, an **informa** business

First edition published 2022

by CRC Press
6000 Broken Sound Parkway NW, Suite 300, Boca Raton, FL 33487-2742

and by CRC Press
2 Park Square, Milton Park, Abingdon, Oxon, OX14 4RN

CRC Press is an imprint of Taylor & Francis Group, LLC

Library of Congress Cataloging-in-Publication Data
A catalog record has been requested for this book

ISBN: 978-0-367-63001-0 (hbk)
ISBN: 978-1-032-13108-5 (pbk)
ISBN: 978-1-003-22769-4 (ebk)

DOI: 10.1201/9781003227694

Typeset in Palatino
by MPS Limited, Dehradun

This book is specially dedicated to my parents, Daniel and Lilian

who have instilled me with a love for Design.

— Eujin Pei

This book is dedicated to my daughter, Hannah Self.

She shows me the way and teaches humility.

— James Self

Contents

Foreword xi
Preface xiii
Acknowledgements xv
Authors xvii
Glossary xix

Chapter 1 Design and design representation 1
    1.1   Design as practice, Design as an activity 1
        1.1.1   Activity and design representation 4
    1.2   Design thinking and design representation 8
    1.3   Industrial design 13
        1.3.1   History of industrial design 15
        1.3.2   The work of industrial design 17
    1.4   Stages of new product development 23
        1.4.1   Concept design 27
        1.4.2   Concept development 30
        1.4.3   Detail design 33
    1.5   Summary 36
    References 37

Chapter 2 Design thinking through representation 41
    2.1   Design representation 41
        2.1.1   Design representation and media of expression 44
        2.1.2   Design representation as construction 45
    2.2   Representation and design cognition 48
        2.2.1   Design representation as reflective-practice 48
        2.2.2   Design representation and ambiguity 51
        2.2.3   Design representation and fidelity 55
    2.3   Representation and design process 58
        2.3.1   Design representation: Concept design 59
        2.3.2   Design representation: Concept development 61
        2.3.3   Design representation: Detail design 63
    2.4   Representations, design problems and solutions 67
    2.5   Design representation: a definition 68
    2.6   Summary 69
    References 70

Chapter 3 Design representation in practice 73
    3.1   The purpose of design representation 73
    3.2   Types of design representations 80
    3.3   Tools of design representations 84
    3.4   Manual 2D media 87
    3.5   Digital 2D media 89
    3.6   Manual 3D media 91

3.7   Digital 3D media 93
3.8   Summary 96
References 97

Chapter 4 Sketches 105
4.1   Sketch representation 105
4.2   Idea Sketch 109
4.3   Study Sketch 110
4.4   Referential Sketch 112
4.5   Memory Sketch 114
4.6   Coded Sketch 115
4.7   Information Sketch 116
4.8   Renderings 117
4.9   Inspiration Sketch 119
4.10   Prescriptive Sketch 121
4.11   Summary 123
References 124

Chapter 5 Drawings 127
5.1   Drawing as design representation 127
5.2   Concept Drawing 129
5.3   Presentation Drawing 130
5.4   Scenarios and storyboards 131
5.5   Diagrammatic Drawing 133
5.6   Single-View Drawing 134
5.7   Multi-View Drawing 136
5.8   General Arrangement Drawing 137
5.9   Technical Drawing 139
5.10   Technical Illustration 140
5.11   Summary 142
References 144

Chapter 6 Models 147
6.1   Models as design representation 147
6.2   3D Sketch Model 150
6.3   Design Development Model 152
6.4   Appearance Model 153
6.5   Functional Concept Model 155
6.6   Concept of Operation Model 156
6.7   Production Concept Model 157
6.8   Assembly Concept Model 159
6.9   Service Concept Model 161
6.10   Summary 164
References 166

Chapter 7 Prototypes 169
7.1   Prototypical design representation 169
7.2   Appearance Prototype 175
7.3   Alpha Prototype 176

7.4   Beta Prototype 178
7.5   Pre-Production Prototype.. 179
7.6   Experimental Prototype 180
7.7   System Prototype 183
7.8   Final Hardware Prototype 184
7.9   Tooling Prototype 186
7.10   Off-Tool Prototype... 187
7.11   Summary 189
       References 191

Chapter 8 **Case studies and conclusions**..**195**
8.1   GMC Case Study 01 195
8.2   Mojavi Case Study 02 200
8.3   Deeptime Case Study 03... 208
8.4   Aero Case Study 04... 212
8.5   Conclusions 219
       References 223

Index...**225**

Foreword

This book responds to the expression 'all you always wanted to know about design representation but didn't know where to ask'. Indeed, the book is a thematic guide to design representation, and the amount of information about design representations it holds is phenomenal. Each chapter is a walk through all aspects of design representation from a particular perspective: design thinking, practice, Sketches, Drawings, Models, Prototypes. Each chapter is an independent unit, with its own references, thereby achieving a somewhat encyclopedic volume that would ideally serve as a textbook. Design representation being the unifying theme of this book, it is chock-full of illustrative images: I counted no less than 121 photographs, drawings, sketches and diagrams. The format of all figures is small, and the authors' policy is to not specify the particulars of the images other than credit to the designers who created them. The intent is to just supply the reader with more and more examples of the type of representation that is discussed in the chapter in question. Despite the independence of each chapter, which causes some unavoidable repetitiveness, there are some most appropriate unifying themes that run through the entire text. The two major ones are the reflective practice nature of design representation, modelled after the theory of Donald Schön, which connects design representation to design thinking, and the ambiguity/fidelity gamut within which design representation takes place. The latter is an interesting way to classify representations by their standing between these two ends of the representational spectrum. Classification is the core of the book's effort: it endeavours to impose order onto design representation by finding the proper place of each instance in a classificatory system. In some cases, the distinction between sub-classes is so fine that it requires an effort to follow the authors' explanation why, for example, a certain model is a functional concept model as opposed to an overall systems prototype. But with practice, it is assumed that the reader will become proficient at such distinctions. Pei and Self wish to support design practice and at the same time they intend to contribute to the study of design cognition. They endorse the claim that designers exercise a particular type of creative cognition, and this cognition is supported by the practice of producing design representations. A careful classification of such representations clarifies which type of representation is appropriate for which phase in the design process: each phase has its own purpose and aim and is best served by a particular type of visual representation: from rough freehand sketches to precise drawings – manual or CAD generated, to Models and then to Prototypes, all the time moving from the ambiguous to the more accurate and detailed fidelity-borne representation. The book celebrates visual representation as used by designers to think with, communicate with, and persuade with. It is a welcome addition to the literature on visual representation and visual thinking in design. It will be very helpful to design students and instructors, and to anyone interested in design thinking and design representation.

Gabriela Goldschmidt
Technion – Israel Institute of Technology
April 2021

Preface

Design thinking is a topic which most designers are familiar with. However, there is still plenty to learn about why design thinking is used or how it can be applied in our practice. As design practitioners ourselves during our early career, we set out to embark on this journey to write a book to demystify this particular type of thinking. We wanted to focus on the use of design representations, having spent a major part of our doctoral degrees investigating this subject. Being full-fledged academics, we see this book as useful and relevant for design students. Within the book, we provide answers to questions students may have wished to ask their tutors but never had the chance about the wide variety of media available, and to understand designerly ways of doing and thinking. Through this book, we hope to share with readers our views and perspectives on practice whether you might be totally new to design or have been a lifelong practitioner yourself. We hope that you will enjoy reading this book. Keep sketching, keep making!

Acknowledgements

This book would not have been possible without the support of our Editor Joseph Clements of CRC Press for his professional advice and assistance in polishing this manuscript. Special thanks to Jonathan Plant of CRC Press who was our first point of contact, we could not have done this without his help and encouragement. We owe an enormous debt of gratitude to Professor Gabriela Goldschmidt who provided the foreword for this book. Finally, we thank many designers, design students and contributors for kindly granting permission to use their figures and images as examples of design representation.

Authors

Dr. Eujin Pei is the Associate Dean (AD-QAA), and Director for the BSc Product Design Engineering programme at Brunel University London in the United Kingdom. He is a Chartered Engineer (CEng), Chartered Technological Product Designer (CTPD) and a Chartered Environmentalist (CEnv). Eujin gained academic experience as a Research Fellow at Loughborough University, Brunel University London and the University of Southampton. He was a Visiting Scientist at Vaal University of Technology and at the Central University of Technology in South Africa. As a Product Design Engineer during the early days of his career, he developed solutions for companies including Motorola, Inc., LM Ericsson, Sennheiser GmbH and Co. KG, and Rentokil Initial of which some products are still manufactured today. Eujin is a Fellow of the Institution of Engineering Designers, Fellow of the Higher Education Academy and Member of the Design Research Society. He is a Fellow and Council Member of the Institution of Engineering Designers, Chair of the British Standards Institute AMT/8 committee for Additive Manufacturing, Convenor of the International Organisation for Standardisation ISO/TC261/WG4 that develops standards and guidelines for Additive Manufacturing Data and Design, and Convenor for TC261/JG67 for Functionally Graded Additive Manufacturing. His work has been exhibited at the Panasonic Centre in Tokyo, Japan; at the NEC Lighting Show in Birmingham, UK; and at the Cooper Hewitt, Smithsonian Design Museum in New York, USA.

Dr. James Andrew Self is a tenured Associate Professor of Design, Director of the Design Thinking Research Lab, UNIST (Ulsan National Institute of Science & Technology), and Visiting Reader, Brunel University London. Prof. Self holds a doctorate in industrial design and worked for several years within the design industry, in London and Sydney, Australia. He holds Associate Editorships and committee positions for a number of international journal and conference publications. Design works and research contributions include international design awards, patents, exhibitions, seminars, workshops and numerous journal, conference and design periodical publications. Design and research works are focused around design thinking, design education and design-driven innovation. Projects attract funding from government bodies and industry sponsors.

Glossary

3D CAD	A three-dimensional model created by computer-aided design tools or software.
Ambiguity	This refers to a condition of being open to more than one interpretation. Design ambiguity is described as being inexact, and in doing provides increased opportunity for interpretation.
Computer-Aided Industrial Design	This refers to the use of dedicated software that generates complex organic shapes and renders photo-realistic images.
Concept Design	This is the first phase of New Product Development that involves generating ideas based on form, function, features, specifications and benchmarking with economic justification.
Concept Development	This is the second stage of New Product Development that involves the selection, development and evaluation of suitable concepts based on set specifications.
Design Representations	Design representations are the means in which designers explore the design problem, identify solution ideas and to arrive at the specification of a clear design intent, usually through the use of Sketches, Drawings, Models and Prototypes.
Design Talk-Back	This refers to the discussion of interaction between representation and the designer's constructed interpretation of a design through reflection (Schön 1986).
Detail Design	This is the third and final stage of New Product Development concerned with aspects of product manufacture, including finalizing the technical description of each component such as materials, colours, finishing, surface texture properties, fits, tolerances, positioning and assembly details.
Design Fidelity	This refers to the degree in which a Design Representation is used to approximate the detail of a design intent, how much information is made available and how prescriptive it is.
Digital	This refers to the use of a system in which information is created, recorded or sent electronically by computers (Longman Dictionary 2005).
Digital Media	This describes electronic forms of media created, viewed and manipulated by computers to produce visual design representations.

Ill-Defined Design Problems	This is sometimes also known as a wicked problem that requires the consideration of many complex variables and through a multi-disciplinary lens. Design problems are seen to be ill-defined as the act of design seeks to arrive at the most appropriate solution, rather than to seek an objectively true or correct answer.
Ill-Defined Solutions	This arises because the final design solution may potentially take an infinite number of forms depending on the way in which the designer both frames the initial design problem and the methods and approaches applied to explore and develop potential solution candidates.
Industrial Design	Industrial Design is focused on aspects experienced by users, including the outlook, usability and the identity of a product.
Manual	This refers to the act of making or working on something with one's hands as opposed to using digital methods (AskOxford 2008).
Manual Media	This describes the use of physical materials and hand equipment to produce a visual design representation through hand-eye coordination and articulation without computers.
Medium/Media	The term 'medium' in singular, or 'media' in plural, refers to tools and materials where something can be expressed, communicated or achieved.
New Product Development	New product development covers the complete process of bringing a new product to market, consisting of Concept Design, Concept Development and Detail Design.
Reflection-in-Action	This is seen as a two-way conversation through which the designer embodies and reflects on design ideas (Schön 1986).
Reflective-Practice	This refers to the interaction between the designer and the representation of design intent through various media, particularly in the use of drawing and sketching.

1

Design and design representation

1.1 Design as practice, Design as an activity

According to Bürdek (2005), design is defined as a plan or scheme devised by a person to develop man-made artefacts for a specific purpose. It can also refer to the arrangement of elements in a product or as a work of art. Design has also been employed to add value to a product and as a communication and retail strategy. From a wider perspective, design brings various elements together to inform the development of appropriate solutions, thereby adding value to people's lives (Pipes 2007). If correctly implemented, good design can enhance the quality of life through innovation, proposing suitable forms, structures and manufacturing requirements that appropriately respond to functional, technical, economic and cultural needs (Fiell and Fiell 2003a). For the purpose of this book, the term *design* is concerned with idea-based disciplines, comprising product and industrial design, engineering design, communication design, architecture, user-interaction/user-experience (UI/UX) design and fashion design among others. We differentiate design from both art and science, and position design representation as critical to the thinking and practices required in an activity aimed at exploring preferred futures.

In contrast to science, where design is justified in being societal, functional and meaningful; science is ruled by formulas, protocol and constraints (Sparke 1983). The natural sciences may be further defined through an objective pursuit of identifiable truths about our external environment (natural sciences). Likewise, the social sciences explore the human condition through the application of the scientific method. In doing so, science builds theory towards an understanding of our world, in the sense that the world is understood through the identification of objective truths waiting to be known. These then manifest as general theory through empirical analysis. Through the scientific method, the resulting theory may be falsified or revised considering new evidence.

Matthew McClumpha's work involving interviewing stakeholders and users, followed by ideation, development and the modelling process is shown in Figures 1.1–1.3.

In contrast to the sciences, design is concerned with creation. In particular, design aims to create potential futures, thereby realising the man-made, material world around us. Thus, design is also pervasive in the sense that everything around us, with the exception of the natural world, has been created by someone (Cross 2021). Although design may draw on scientific understanding, design does not aim to identify objective truths. Rather, design proposes interventions that have the potential to improve the existing situation through the creative proporsition of potential futures, which then may be implemented through various production methods and processes.

Although the arts and design share common ground, in particular their role as reflections on culture and attitudes and relation to societal trends, unlike the arts, design's

DOI: 10.1201/9781003227694-1

FIGURE 1.1
User studies (McClumpha 2020).

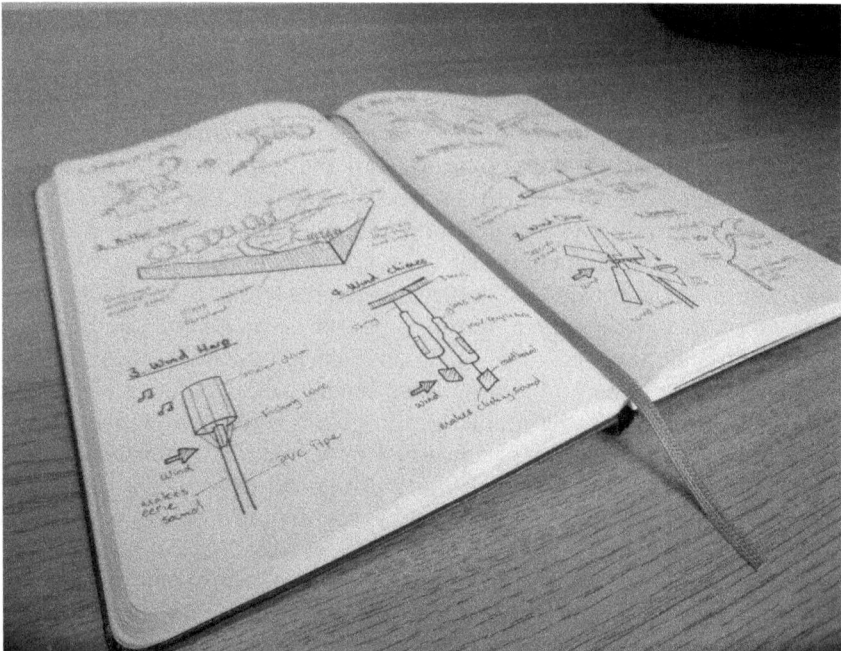

FIGURE 1.2
Photograph of a designer's sketchbook (McClumpha, 2020).

FIGURE 1.3
Functional Concept Models being tested at a beach (McClumpha, 2020).

concern extends beyond the representation of the world, culture, politics and values. Although design may very well embody these broader themes, messages and concepts. Instead, design has a greater focus on how interventions may add value for *others*. This *user-focus* often requires an understanding of and consideration for both emotional aspects (*design aesthetics, design semantics*) and functional requirements (related to end users). As a result, design must provide interventions that are appropriate for and add value to the lives of others on both emotional and practical dimensions. In this more holistic sense, design often requires a kind of thinking that can support a synthesis of many issues when exploring, defining and creating design solutions.

If we frame design as the proposition and implementation of more appropriate futures, the act of designing involves creatively building the nature, appearance and social function of objects (Tjalve 1979). It entails the use of problem-solving methods and creativity to produce the desired properties of a product (Andreasen et al. 1988). As design ideas are formulated in the mind, various elements and constraints are considered, balancing practical function with aesthetics (Cross 1996). However, when mental images in the mind (internal representations) are produced and externalised through sketching, drawing and modelling (Goel 1995), the externalised representation of intent act as distributors of cognition to become part of the information used to generate the next idea, or to develop the current concept. These simultaneous representations assist and support the mental sorting of information, allowing the designer to consider many other factors required in the process of design (Tovey 1989).

In terms of process, design can also be considered an act of problem solving through trial and error that involves iterations (Roozenburg and Eekels 1995). These steps are repeated as no firm decisions are usually made during the first round of iteration. These iterations occur throughout the design process and involve innovation, analysis, decision-making and evaluation. During this process of iteration, an approximate solution to an initial design problem is worked through, and then fed back into the process for an improved solution. This continues until the desired solution is achieved when more information becomes available, and consequently ceases when all available resources are depleted (Wright 1998). Cross (1984) suggested that design is an open-ended and ill-structured process, having no clear solution at its comencement. Answers cannot be obtained through the use of formulas since the goals, constraints and criteria may be poorly understood in the beginning and often change and evolve as the design progresses. In addition, formulating the problem is difficult because in design, there are no true or false answers. Instead, problems and their solutions are often assessed as being either good or bad, or more or less appropriate.

Design problems therefore are often ill-defined (Dym and Little 2009), whereby an approach is needed to counter the instability of the situation. This is an effect of the ill-structured design problem, and the resulting need to deal with a myriad of potential avenues of development towards possible solutions. This particular moment of the design process requires the freedom for creativity to take place as the designer applies past experience of similar problems and related processes to structure the problem. To support this structuring, the designer must explore ideas through solution propositions (Stempfle and Badke-Schaub 2002). In light of ill-defined design problems, designers attempt to identify and address potential solutions, through propositions that use design approximations as Sketches, Drawings and other representations of design intent. These representations are employed to help define and structure problem-solution pairs. Through representation, the design moves to converge on a matching problem-solution pair as a propositional answer expressed as a solution candidate (Cross 2021). Within the process of structuring and pairing, an ability to distribute cognition through the expression of ideas represented through Sketches, Drawings, Models and Prototypes appears critical to the identification, evaluation and further development of solution candidates.

1.1.1 *Activity and design representation*

Because the designer must attempt to match a design problem to its solution by evaluating the appropriateness of the design idea, the process of design practice can be described as being solution-focused and goal-oriented (Vissler 2006a; Press and Cooper 2003). It is, in part, due to the ill-defined nature of the design problem and the resulting search for solution ideas, that the embodiment of intentionality through Sketches, Drawings, Models and Prototypes as representations of intent appears critical to design activity. These design representations act as scaffold through which designers are able to explore problems, identify matching solution candidates and eventually arrive at the specification of a clear intention (Vissler 2006b, p. 116; Goel 1995 p. 91). In this sense, design representations are employed as a means to support designerly thinking between ill-defined problems and their solutions. In this supporting role, representations appear to exist as extensions of thought, rather than being mere expressions of internal representations in the mind.

As extensions of thought during design activity, Schön (1983), in his seminal work on reflective practice, refers to the designer's relationship with design representation through the metaphor of conversation between the designer and design representation. These representations 'talk back' and 'back-talk' as the designer reflects on the implications of design solution approximations. The complexity of design activity, with its competing influences and responsibilities, requires the testing of design intentions as, 'the designer shapes the situation in accordance with an initial appreciation of it, the situation *"talks back"* and the designer responds to the situation's *"back-talk"'* (Schön 1983, p. 79). Like Schön (ibid), Lawson (2006) describes the designer's reflection through representation using the metaphoric analogy of engaging in 'a conversation with the design activity' (ibid, p. 265). Goldschmidt (1994) describes design activity as a form of 'visual design thinking' (Goldschmidt 1994, p. 160), where expression of design intent interacts with thinking to provide opportunities for understanding the potential of design ideas suitable as solutioncandidates. This is shown in Figures 1.4 and 1.5. Sam Gwilt produced variations of his concepts to explore different product forms and shapes. He uses both digital and physical representations to help him decide which of these ideas would be the most suitable.

Existing studies offer a definition of design activity that focuses on the construction of and reflection upon design intent through representation. The physical embodiment of the designer's own thoughts mediated through the use of design representations provideopportunities for the exploration and evaluation of solution ideas that would otherwise be impossible. This then suggests the critical importance of representation for design activity.

As discussed previously, reflection upon representation provides the opportunity to iteratively identify a matching problem-solution pair. During this identification process, the practitioner will reflect on and re-evaluate design intentions (Schön 1983). These reflections provide momentum for the designer to assess and evolve the design intent, in which 'there is a continually evolving system of implications within which the designer reflects-in-action' (Schön 1983, p. 103). Design representation, then, is identified as the means through which design activity must proceed if a more appropriate solution is to be identified. This is also represented by Gupta and Murthy (1980) in Figure 1.6 who claimed that the iterative cycle of design requires a process of innovation, analysis, decision-making and evaluation.

FIGURE 1.4
Reviewing a Single-View Drawing with Idea Sketches in the background (Gwilt, n.d.).

FIGURE 1.5
Reviewing design representations as Study Sketches (Gwilt, n.d.).

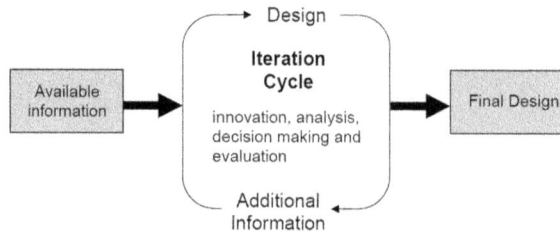

FIGURE 1.6
Design as an iterative process (Gupta and Murthy, 1980).

Returning to the analogy of reflection-in-action (Schön 1983) as a two-way conversation through which the designer embodies and reflects design ideas, Lawson (op cit.) and Purcell and Gero (1998) explore the ways in which these conversations inform an exploration of the design problem through the tangible embodiment of solution ideas as Sketches, as well as the use of Three-Dimensional Computer-Aided Design (3D CAD). His discussion of design embodiment through 3D CAD indicates the influential role that the media or tool of expression has upon reflection. For example, Lawson (op cit.) describes the ways in which 3D CAD may constrain a conversation as, 'a halting clumsy process that more closely resembles the assembly of a sentence in a foreign language' (Lawson 2004, p. 70). If the tool of representation can influence outcome, this insight indicates how design representation and reflection are critical aspects of what it means to engage design activity. If a tool's particular expression of intent due to representational constraints influences the final solution outcome, then this suggests a relationship between thinking and representation.

Dorta et al. (2008) point to the reflective nature of design activity with a focus on the influence of design representation in terms of how representations relate to the changing requirements of the development process. Dorta et al. (ibid) further describe how designers use externalised Models of proposed design solutions to interact with internal, mental images. This interaction then informs a process whereby the externalised embodiment of ideas through design representation results in unexpected or unforeseen insights. In this sense, Dorta et al.'s (op cit.) discussion of interaction between representation and the designer's constructed interpretation of a design through reflection, relate to Schön's (1983) notion of *'talk-backs'*, (p. 79). As shown in Figure 1.7, Jinghua Li produced Sketches of his bedside clock, in which he would also need to consider the user interface of the display. This required him to negotiate between the form of the product display, as well as externalising how the information on the screen would appear to the user.

These studies evidence the relationship between the externalisation of design ideas as representations of intent and thinking towards design problems and their paired solutions. Although reflective practice as a paradigm for understanding design activity provides a broad indication of this relationship, there is still little theory or conceptual modelling to scaffold a universal understanding of design representation. For example, less is said about the underlying cognitive, perceptual or psychological mechanisms that may underpin relationships between design representation, the designer's reflection upon them and how this interaction leads to the identification of a problem-solution pair as part of an often iterative design activity. One way into developing an operational model of design representation's use in practice may be through its contextualization as part of *design thinking*.

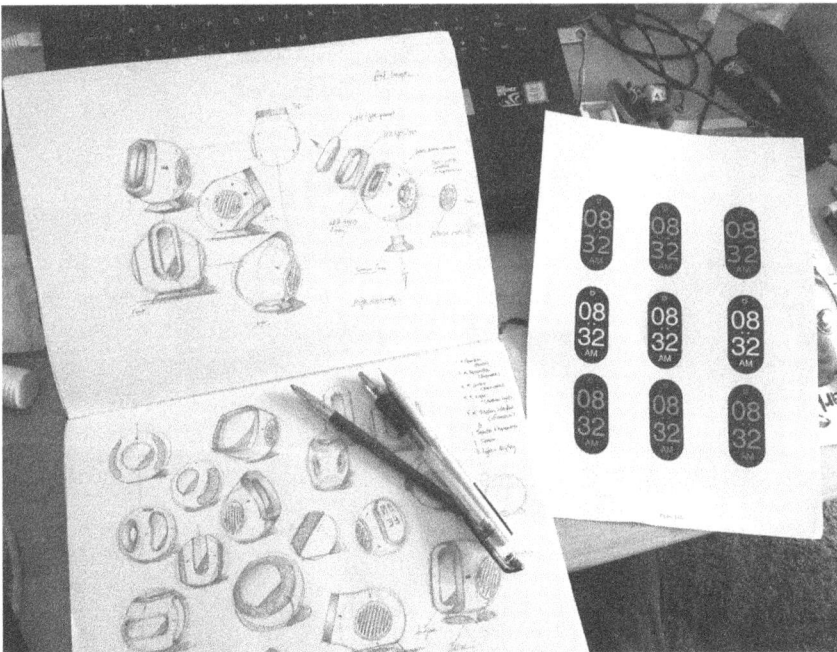

FIGURE 1.7
Idea Sketches and Study Sketches (Li, n.d.).

1.2 Design thinking and design representation

One of the first generations of design models was established by Rittel and Webber (1973) who proposed that the design process could be conceptually identified as a series of discrete steps – *understanding, collecting, analysing, developing, assessing* and finally *testing* of the solution (Erlhoff 1987; Bousbaci 2008). Other scholars contributed to a more explicit understanding of a process of design with a morphology proposed by Asimov (1962); a formalised design checklist proposed by Archer (1965); a framework for evaluating design solutions by John R. M Alger and Carl V. Hay, and the introduction of the *new product development* process which was first described by Jones (1969) (Figure 1.8) and then later expanded and improved by Cross (2021).

According to Urban and Hauser (1993), the design process may be conceptualised through a linear series of discrete steps or activities, including *idea generation, product development* and *product commercialisation*. Other models of these activities include those by Tjalve et al. (1979), French (1985) (Figure 1.9) and Pahl and Beitz (1996) (Figure 1.10). Common among these models is their structure of beginning with an initial statement of

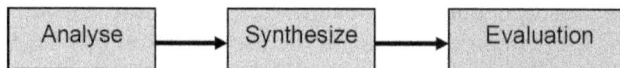

FIGURE 1.8
The design process according to Jones (1969).

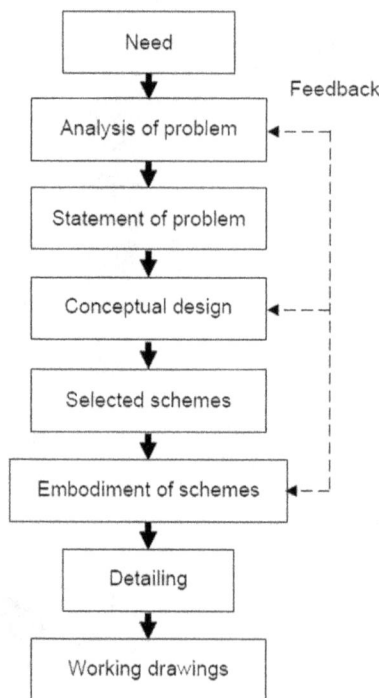

FIGURE 1.9
French's (1985) block diagram of the design process.

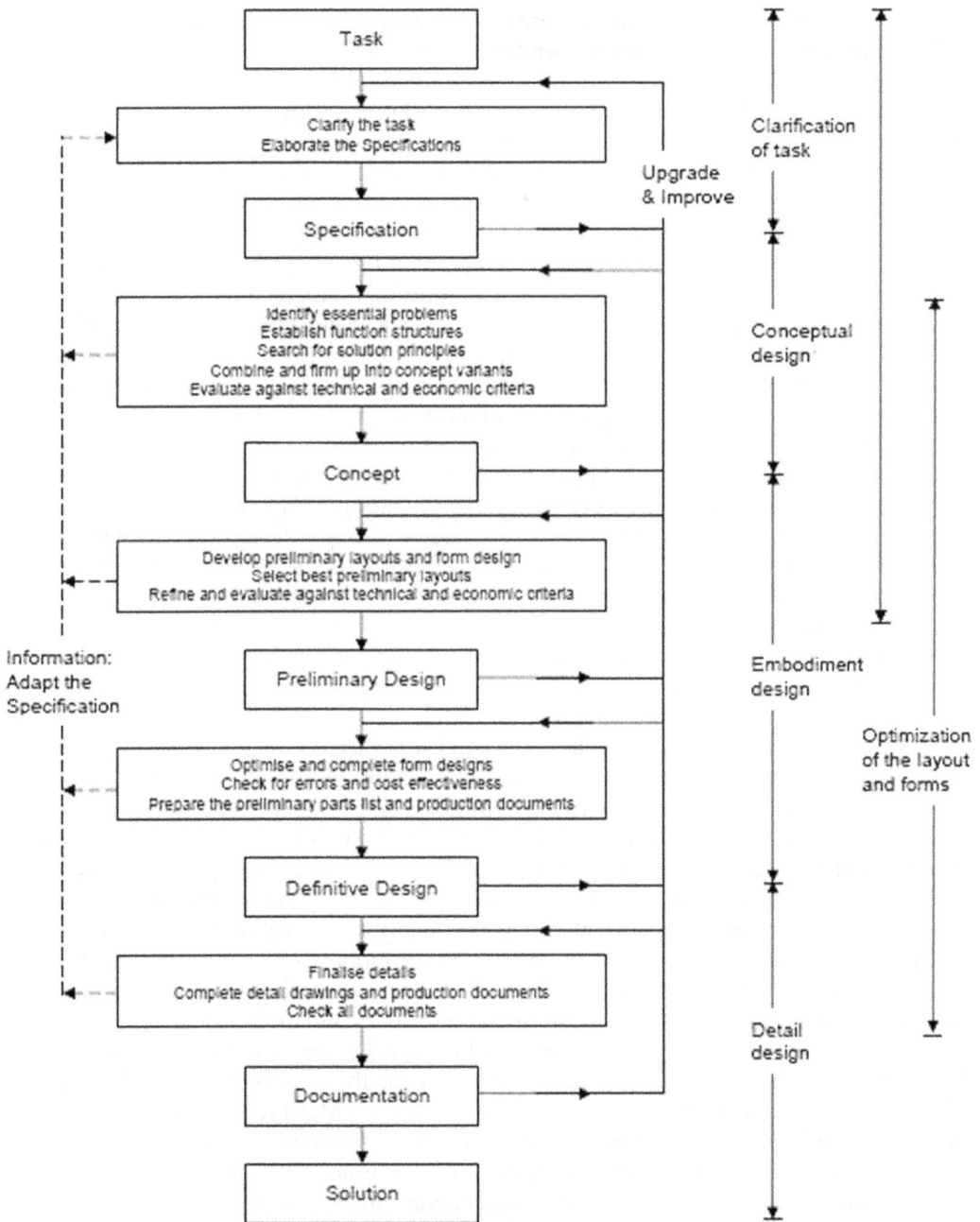

FIGURE 1.10
Pahl and Beitz's (1996) phases of the design process.

needs and problem analysis. A *conceptual design* phase then generates and selects a design idea. Next, an *embodiment* stage allows the chosen concepts to be worked further. The last phase of *detail design* enables many small but essential points to be made. At each of these discrete phases in process, design representations are used to express and communicate intent at various levels of detail and fidelity. In fact, these models indicate a linear relationship between representational fidelity and the phase of the process. As the design process moves from concept, through embodiment and into detail design, the fidelity of design representation increases to provide more detail at higher levels of resolution.

Related to a need for different types of representation as dependent on the phase in process, design at an individual level requires conveying information clearly to others. It also requires personal characteristics such as flair, ability, intuition, creativity, judgment, communication, reflection, feeling and experience (Schön 1983). To aid this, Pahl and Beitz (1996) proposed guidelines such as using a suitable style; having a structured and unified form; good use of colours and complementary graphics. Design also occurs among individuals as a shared inquiry and a continuing dialogue among a broad circle of active stakeholders (Hack and Canto 1984). Design is a social activity where people from different backgrounds build a shared vision when working together. This social process involves negotiation and building consensus, bringing the perspectives of individuals together to collectively develop the final product (Bucciarelli 1994). However, as different stakeholders have competing and conflicting objectives, the act of design becomes much more complicated (Sebastian 2005). At a macro level, the expression of intent through representation is dependent on both phases in the design processes and how various stakeholders see and understand ideas as expressed through design representation.

How then do designers and other stakeholders make sense of design expressed through various representations? Within the field of design, design thinking has more recently emerged as concerning visual thinking emerged as concerning visual thinking whereby a mental image of an object is made visible through cognitive processes such as perception, imagination and communication (Persson 2002). According to Rodriguez (1992), the visualisation process is an important ability for designers, classifying visualisations as those that can be seen, those imagined in the mind and those drawn or modelled in a physical form. When constructing visual images, the designer introduces a myriad of features including form, proportions, orientation, material, colour, symmetry, contrast, repetition, and so on. These visualisations of design intent are an important part of design thinking. That is, the embodiment of solution ideas as design representations appear to provide a necessary distribution of cognition between designer and design representation. This then allows the inclusion of multiple features when thinking towards potential solution candidates.

Although sensation and perception involve multi-modalities of sensory information, research in design thinking suggests that the visual system is the primary stimuli (Kosslyn 1994). Seeing, imagining, reflecting and drawing are all interrelated, where past experience and hard-wired interpretation of the world filters what we see, and seeing then stimulates our imagination which in turn interacts the design representation such as a Drawing or illustration (Dorta 2005). This is in line with McKim (1980) who established that visual thinking involves the interaction of mental (imagining), graphical (drawing) and perceptual (seeing) processes. McKim's (ibid) discussion of visual thinking in design also indicates the distributed nature of design cognition. Design representations are used as a scaffold to aid the proposition, development, and communication of potential solution candidates, and as a means of visual thinking to understand the potential of design ideas. The photographs in Figure 1.11 and 1.12 show Ajharul Choudhury who produced several iterations of his concept and obtained feedback from potential users to discuss further design possibilities and to confirm the users' preference.

Thinking styles can also be classified into left and right hemispherical use of the human brain. Evidence from existing research on cognition has shown that facts, numbers and words are more greatly associated with the left brain; and aesthetics and creativity involve the right hemisphere of the brain (Burghardt 1999). The left side of the hemisphere in the brain is logical

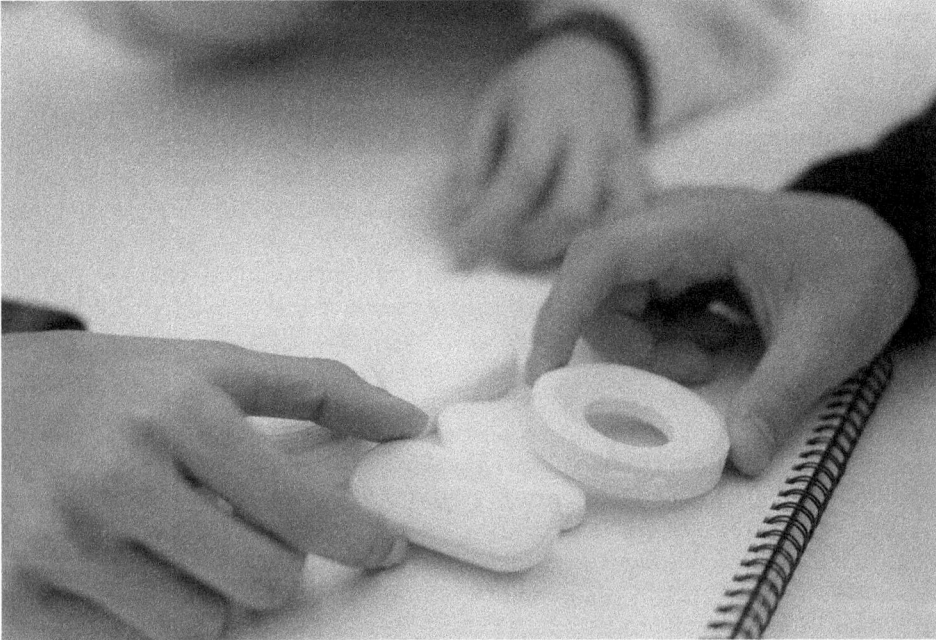

FIGURE 1.11
Comparing design ideas using Appearance Models (Choudhury, n.d.).

FIGURE 1.12
Testing Appearance Models with potential users (Choudhury, n.d.).

and systematic (Jones 1992). Information processing is serialised to investigate deep into a problem space with careful decisions at each stage. This rational, verbal and analytic thinking is known as serial thinking (Cross 2021). In contrast, the right side of the brain's hemisphere generates more alternative ideas and visuals that are associated with lateral thinking, seeking as many choices as possible and doing things out of sequence (Bradshaw and Nettleton 1983). This is also known within the design research community as holistic thinking (Cross 2021) and the vertical approach of the left hemisphere ensures that the individual evaluates information logically and objectively. It is analytical, judgemental, critical and selective. In contrast, the lateral thinking of the right hemisphere is random, simultaneous and generative where the individual thinks in several broad directions by combining bits of information into new patterns and expands possibilities into new ideas. The key difference is that vertical thinking regards an idea as a converging goal, whereas the goal of lateral thinking is to generate as many divergent ideas as possible (Tovey 1984; Shetty 2002).

Applying the aforementioned theory on cognition to design activity, left-brain convergent thinking may be most useful in narrowing a design space by filtering the best alternatives with logical and structured methods. In contrast, right-brain divergent thinking is expansive, seeking to identify a greater number of ideas and choices by thinking 'outside the box' (Dym and Little 2009). This is shown in Figure 1.13 where Tovey (1984) discussed aspects of left and right brain thinking approaches and how they are entwined in the design process.

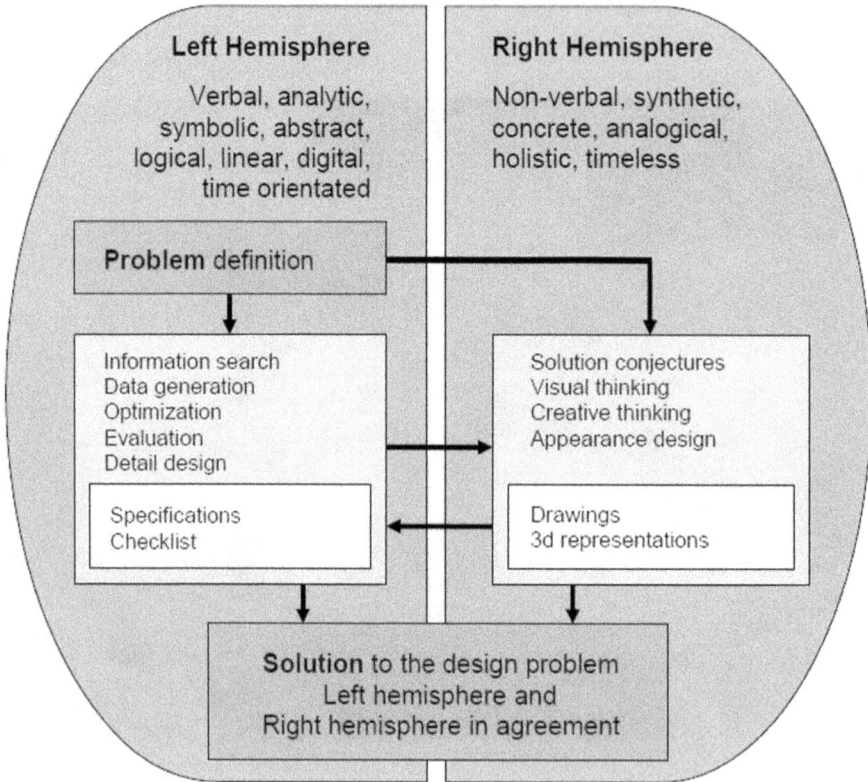

FIGURE 1.13
Tovey's (1984) dual processing model of design.

Left Hemisphere

Facts, Numbers, Words
Logical, systematic, Rational
Sequential
Verbal
Analytic
Serialist
Vertical
Linear processing
Convergent
Narrows and filters
Logical / Linear / Digital
Time orientated

Right Hemisphere

Ideas and Visuals
Out of sequence
Random
Non-verbal
Synthetic
Holistic
Lateral
Simultaneous processing
Divergent
Expansive
Intuitive / Spatial
Timeless / Diffuse

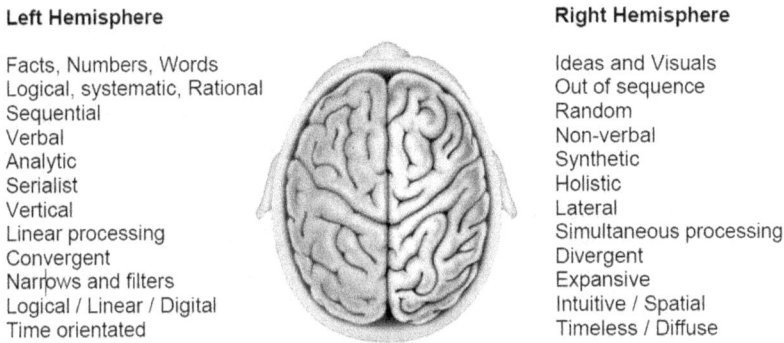

FIGURE 1.14
Cross' (2000) left and right hemispheres of the brain.

Divergent thinking is also associated with creativity where the use of brainstorming activities and ideation generates a large number of Drawings and Sketches that allow members to explore beyond conventional ideas (Eissen and Steur 2008). Cross (1983) suggested that these two approaches can be summarised where the design process often begins with a divergent manner and then converging as the possible solutions are filtered down into a well-defined, single solution. Design representations may support divergent thinking during the design ideation phase to seek a variety of solutions. Divergent thinking is also associated with creativity where the use of brainstorming activities and ideation generates many Drawings and Sketches that allows members to explore beyond conventional ideas (Eissen and Steur 2008). Cross (1983) suggested that these two approaches can be summarised where the design process often begins with a divergent manner and then converging as the possible solutions are filtered down into a well-defined, single solution (Figure 1.14). Design representations may support divergent thinking during the design ideation phase to seek a variety of solutions.

The types of representation used (low fidelity to high detail) also appear to relate to the kinds of design thinking required at different stages of the design process, from divergent exploration to convergent specification. It may be that convergent and divergent thinking through design representation will influence the type of representation used as means to distribute thoughts between the mind and expressing the representation. Lateral thinking is required during concept design being reflected in more ambiguous and open representations used (i.e. concept sketching). In contrast, structured and detailed specifications through Engineering Drawings support convergent thinking during detail design. If this is the case, it provides further evidence to indicate a critical relationship between design thinking and design representation.

1.3 Industrial design

To understand how design representation relates to and is informed by a process of new product development, we will first clarify the background of the discipline of industrial designers. According to the Dictionary of Art Terms (2003), industrial design is the reasoned application of aesthetic and practical criteria for the design of machine-made artefacts, with the hope of creating a successful marriage between aesthetics and functionality. Goldschmidt (1995) further acknowledges that the field of industrial design lies between engineering and other artistic design disciplines, with an expressed purpose of creating artefacts that deliver or apply advancements in engineering and science. For example, the telephone is widely acknowledged as an invention of Alexander Graham Bell, but it was the industrial designer who gave the phone its form (Hannah 2004). In this sense, and beyond the engineered aspects of industrial design, well-designed products provide feelings towards aesthetic and emotional experience through their use (Billings 2006). By providing an aesthetic experience that moves beyond functionality, manufacturers are able to increase their competitive advantage, making products both usable and emotionally attractive for consumers (Ashford 1969; Bohemia 2002). In addition, industrial design can be used to communicate the manufacturer's image and promote the integrity of the product to enhance sales (Yamamoto and Lambert 1994).

The goal of the industrial designer is to understand and achieve the requirements of both user and manufacturer (Holme 1934). The industrial designer plans and creates physical artefacts suitable for mass production by synthesising engineering, technology, materials, and aesthetics, balancing the needs of users from technical and social perspectives (Heskett 1980; Gemsera and Leenders 2001; Fiell and Fiell 2003b). Apart from aesthetics, the industrial designer should also possess a working knowledge of manufacturing methods, human factors and sustainability. According to Tovey (1997), the role of the industrial designer is to design products and product experiences that incorporate market, user and engineering requirements. Core to an industrial designer's skill set is an ability to visualise the product concept and represent alternative design solutions. Chris Hill's work in Figure 1.15 shows some physical Prototypes being produced for the consumer market.

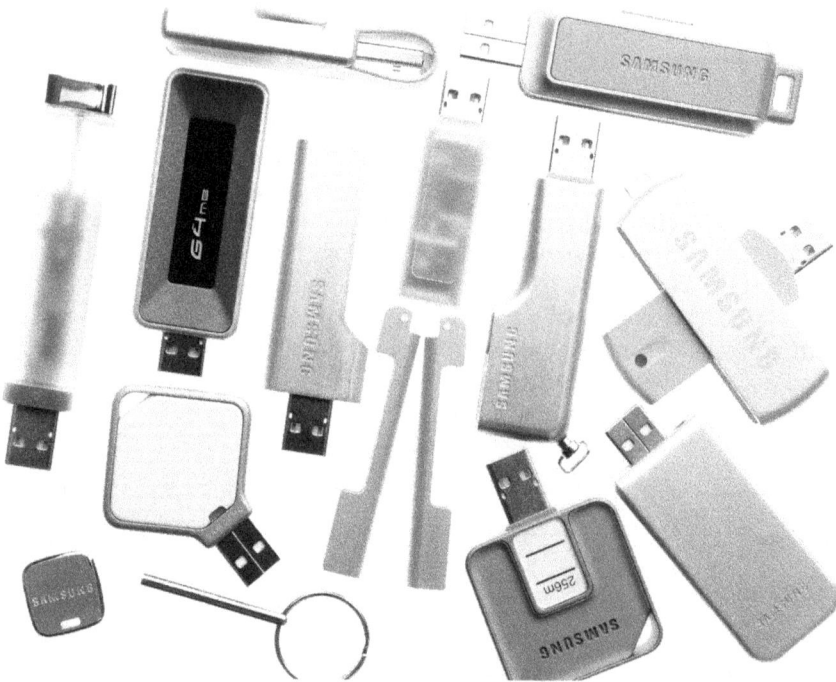

FIGURE 1.15
Chris Hill USB. CMF Design Models (Hill, n.d.; Samsung Design Europe Studio).

The *Industrial Designers Society of America* (IDSA) refers to industrial design as the professional service of creating and developing concepts and specifications that optimise function, value and appearance of products and systems for the mutual benefit of both users and manufacturers (IDSA 2006). In some literature, the terms product design and industrial design may be used interchangeably. However, the term 'product design' is often used to refer to products and this has been felt to be too limiting. To add to the confusion, an early report by Corfield (1979) defined product design to comprise both engineering design and industrial design. Even today, the *Design Council* in the UK uses both product and industrial design terms interchangeably when describing product creation activities. To achieve consistency throughout our discussion of design's relationship to representation, the term industrial design will be used to position our focus on the design of products for industrial manufacture.

1.3.1 *History of industrial design*

Industrial design has a young history, with its roots stemming from the *Arts and Crafts* movement and the Bauhaus school of design in Europe. The term 'industrial' is used because products are manufactured by industrial processes (Hirdina 1998). Aesthetic design has long existed since ancient civilisations with products such as Greek pots, Byzantine ornaments, and artefacts in Egyptian temples. For centuries, objects were created by craftsmen who planned and produced artefacts from start to finish. It was in the early 19th century that witnessed the industrial revolution where mechanical production and a divided system of labour superseded the use of hand-production (Heskett 1980). The difference between the craftsman and an industrial designer is that the craftsman planned and created the product, while the industrial designer does not produce the product (Sparke 1983). Consequently, when an industrial designer designs a one-off product, the term 'industrial' is dropped and they are acknowledged as a 'craft designer' (Campbell et al. 2006). The outbreak of the First World War saw the implementation of standardised and mechanised production, with very little emphasis on aesthetics. In the 1930s, a saturated market and the Great Depression in the USA made manufacturers realise that they could boost revenue and seek a competitive advantage by improving the appearance of products. These visually trained individuals were tasked with making products irresistible and thereby filling the gap between art and manufacture (Woodham 1983). An example is shown in Wai Lim Chan's work for Dell monitors, portraying these devices to be beautifully designed with attention to aesthetics and detail (Figure 1.16).

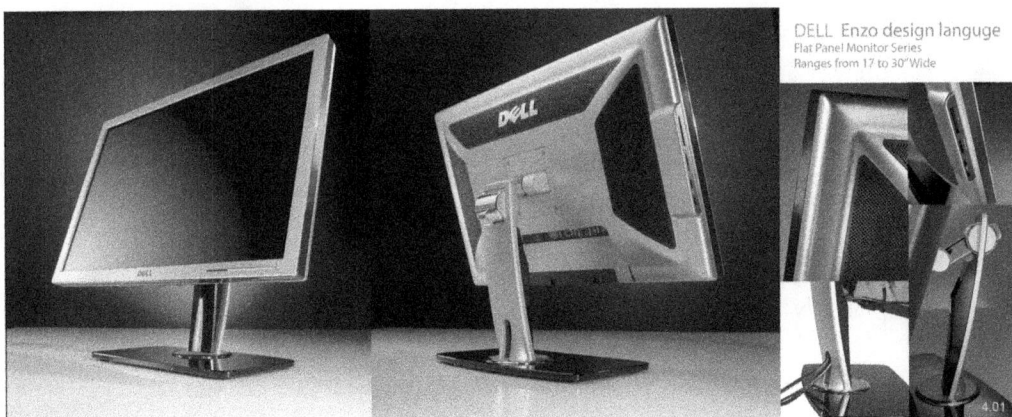

FIGURE 1.16
DELL Enzo design language expressed as Renderings (Chan, n.d.).

As an example of the emergence of the profession of industrial design, Peter Behrens has been regarded by many as one of the pioneering professional industrial designers. Originally an architect, Behrens was engaged by AEG as an artistic advisor to enhance the company's products. Behrens worked by varying finishing, form and proportions based on a standard component. This approach made his work novel, distinguishing himself as one of the first modern industrial designers (Heskett 1980). Since then, industrial designers have extended their responsibilities to include market research, trend analysis, ergonomics and usability studies. The key activities of modern industrial design also include innovating and developing conceptual design ideas. To do so, the industrial designer must be adept in externalising thoughts, to communicate and sell the idea to the client (Pipes 1990). In addition, the industrial designer should be skilled in visual design representations, from creating quick, simple Sketches to producing highly detailed Prototypes that are essential to communicate the design idea (Garner 1999). Chris Hill's work in Figure 1.17 shows a reference level model of a handheld PDA for Samsung Design Europe Studio. This includes the user interface display in which this highly detailed prototype seeks to communicate the final look and feel of the intended product.

Tovey (1989, 1997) suggests the industrial designer has a particular concern towards the appearance of products and in representing design concepts. Designers should also possess a sound grasp of the market, the user's needs, engineering requirements, and manufacturing limitations. In terms of aesthetics, the industrial designer should provide the product with a sense of unity, coherence and individuality to produce a distinct

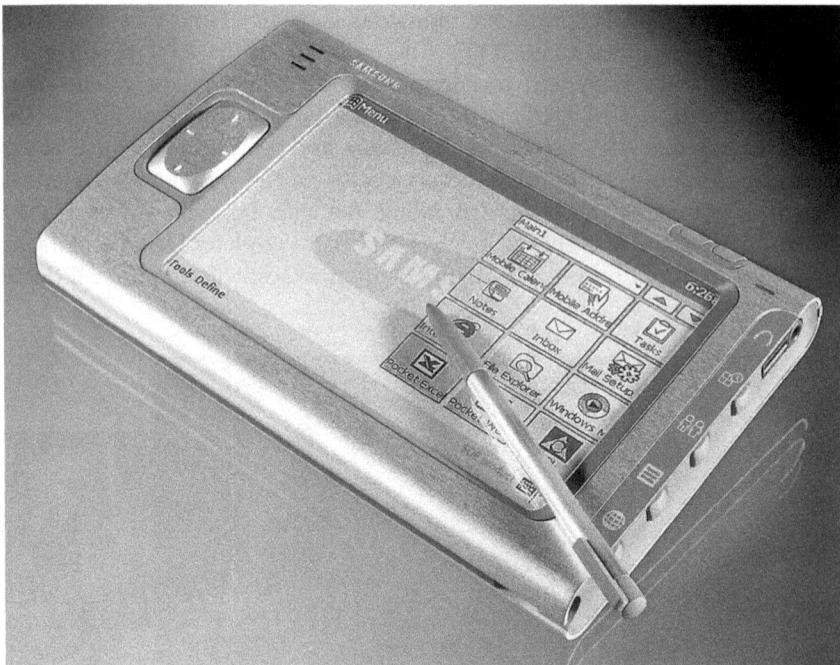

FIGURE 1.17
Mock-up of the Samsung Airpen concept as an Appearance Prototype (Hill, n.d.; Samsung Design Europe Studio).

product personality. An example is the German company Braun that has developed products such as shavers by jointly working between designers, engineers and marketing experts, combining technological innovation with clear aesthetic expression, to create products that are distinctive, desirable, functional and beautiful (Fiell and Fiell 2003b).

Good designs, then, aim to provide an optimal balance between functional, emotional, aesthetic, manufacture and the ethical needs of the consumer. Efficiency, economy, sutainability and ease of maintenance must also be considered (Kristensen 1995). As manufacturing becomes even more advanced, industrial designers are expected to be proficient in the use of CAD, and to work with various disciplines to develop increasingly complex products (Hannah 2004). Persson (2005) explains that the industrial designer's work is focused on aspects experienced by users, including the outlook, usability and identity of a product. Industrial designers work by first creating an overall solution and then work on the details (Tovey 1997).

1.3.2 *The work of industrial design*

Industrial designers use representational codes that take the form of Sketches and Drawings, considered to be the most convenient form of articulating design ideas (Robertson 1996; Kavakli et al. 1998; Verstijnen et al. 1998). Elements of shape, colour, material and texture are all outputs from an industrial designer's work when developing products. The exploration of form is one of the most influential characteristics in the development of products and designers often focus on the form of the product early in the process. Representation of form through Sketches allow shapes to be externalised and communicated, as well as to develop design detail (Figures 1.18 and 1.19).

Design representation through sketching assists in the recording and reporting of the designer's cognition and provides a self-reflective conversation as unexpected shapes emerge (Garner 2006). However, the construction of meaning from the sketching process is also influenced by the designer's background and field of expertise. For example, in an earlier study, Wang and Shen (2002) identified differences in the perception of form between architects and industrial designers. Architects often identify emergent shapes associated with transformational processes, while industrial designers tended to interpret

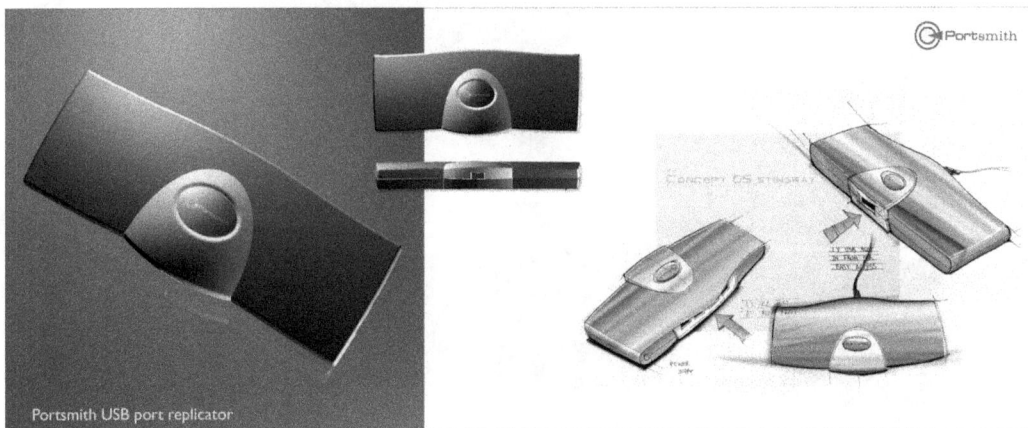

FIGURE 1.18
An example of a USB port in the form of a Rendering and an Information Sketch (Chan, n.d.).

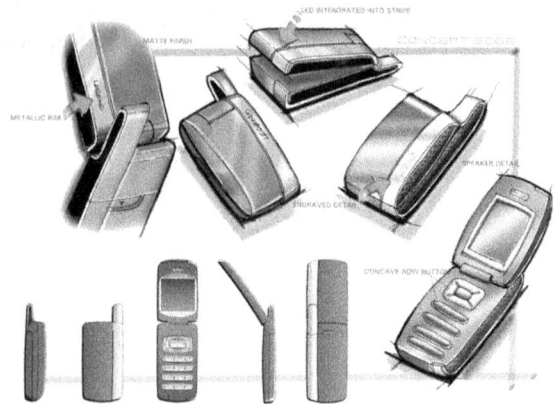

FIGURE 1.19
An example of a mobile phone represented a Rendering and Information Sketches (Chan, n.d.).

volumetric sub-shapes from 2D representations. The professional industrial designer should be skilled in communicating how the final product should look (product form) and to ensure that the design intent is accurately conveyed (Cross 2007). To achieve these communicative requirements, design representation of various degrees of detail and fidelity are produced (Figures 1.20 and 1.21).

It must be noted that the simplicity and spontaneity of representations such as Sketches should not be restricted only to 2D and on paper. Where form and surfaces need to be further explored, industrial designers may use physical materials to create 3D forms, more popularly recognised as the act of 3D sketching or sketch modelling. Also, 2D representations lack the tactile experience and do not provide confidence for stakeholders to proceed directly for

FIGURE 1.20
Physical mock-ups as Appearance Models that express the design intent in terms of colour and finish. Volume and ergonomics were an important consideration in this case (Hill, 2012) (ID By the-Division).

FIGURE 1.21
Concept Renders to inform final model finishes (Hill, 2012) (ID by the-Division).

manufacture. This justifies the need to produce a non-working Block Model or a Working Prototype as a close representation of the final product (Evans and Wormald 1993). The 2D representations allow ideas to be seen and tested in a tangible and low-cost way (Frishberg 2006). Consequently, the delivery of a final prototype signifies that the input of industrial design is decreasing, with subsequent follow-ups limited to fine detailing, manufacturing, testing and post-production support.

Apart from creating physical representations, Computer-Aided Industrial Design (CAID) has also gained importance due to increased ease of modelling, manipulation and visualisation of 3D surfaces (Evans and Wormald 1993). Digital design methods allow information to be sent directly to the manufacturer for production, thus saving time. The process of using pencil and pen Sketches and then moving into solid Models with the use of CAD/Computer-Aided Manufacturing (CAM), and rapid prototyping technologies such as Additive Manufacturing (also known as 3D Printing), is a popular approach among industrial designers (Utterback and Vedin 2006).

In particular, rapid prototyping through the use of 3D printers is now commonly used during the design process to produce variations of an idea, to study complex forms, mechanical structures, or to confirm the finalised design of a product (Figure 1.22). Rapid prototyping may be further classified into three categories of tools where the first are 2D cutting devices such as vinyl and laser cutters, second are subtractive milling machines that carve foam or other softer materials and third are additive manufacturing machines that build solid Models from extruded material, binding or fusing of loose powder, laminating sheets of material or curing liquefied resin. When using rapid prototyping, it is important to consider the time required to create the digital Model and the time it takes

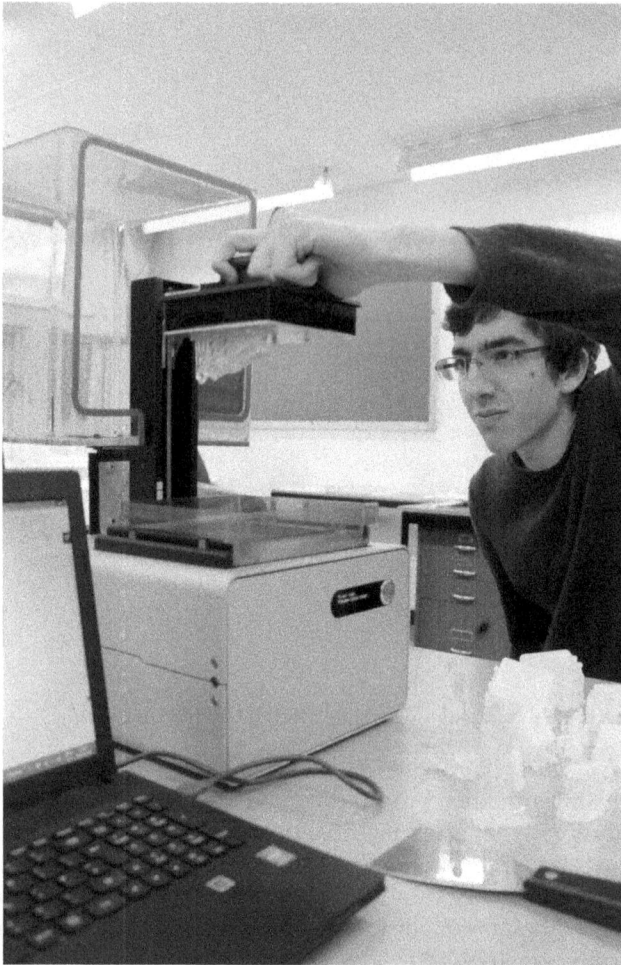

FIGURE 1.22
A Vat Photopolymerisation Additive Manufacturing machine (3D Printer) (Aligizakis, n.d.).

for the manufacturing process. These technologies should therefore be selectively applied as opposed to the use of manual methods.

When asked about one's design approach, principal industrial designer, Mario Turchi of ION Design suggests that the moment begins by thinking and looking for references and then forming ideas by sketching them on paper. A meeting then takes place to bring the project members together for discussion. The design team goes back and returns to present the developed ideas. After more brainstorming, the initial design concept is born (Hannah 2004). Other industrial designers prefer to adopt a more hands-on approach, such as Mark Lim of Conair Corporation in Connecticut who describes his work as involving study and research, sketching forms and creating 3D CAD Models. Other industrial designers prefer a more holistic thinking approach, such as Tucker Viemeister of Springtime-USA who de-scribes a careful analysis of problems, investigating for improvements to daily life, looking for added features, finding applications for technology, and even dreaming of ideas (Pipes 2007). In terms of corporate working approaches, generally, most European industrial design consultancies work up to the General Arrangement Drawing stage and are rarely involved with the technical details of manufacturing or testing. However, British and American companies ensure that their designs are seen right through to production, certifying that the original design intent has been retained (ibid). Most large corporations have an internal industrial design department and small companies usually contract design services from consultancies. In all cases, industrial designers are always required to work with other disciplines including engineering designers, to generate, develop and evaluate concepts for the product throughout the stages of new product development as shown in Figures 1.23 and 1.24 where teamwork is an essential skill.

In describing the contribution of the industrial designer at each stage, Ulrich and Eppinger (2003) indicate that at the concept design stage, the industrial designer conceptualises the product in terms of the overall form and the user interface. This is usually done by means of quick and simple Sketches that provide a fast and cheap way to express ideas. Sketches act as a facilitator and to record creative acts, providing further opportunities for improved evaluation, restating and clarifying the problems (Temple 1994). The concepts are then evaluated by the design team against customer needs, technical feasibility, cost, and manufacturing considerations. In the concept development stage, the use of Models helps industrial designers to express and visualise product concepts in a tangible form and are presented to the stakeholders and customers to gain feedback. At the embodiment design stage, industrial designers translate the Sketches and Drawings into Models with further technical details (Figures 1.25 and 1.26). Realistic renderings are used to convey realism about the product's features and its functionality. This includes the selection of colour, textures and material finish. The final stages see the delivery of the documents necessary for the manufacture of the product and may include General Arrangement Drawings and Working Prototypes (Ulrich and Eppinger 2003).

FIGURE 1.23
An ideation workshop in progress (Chan, n.d.).

FIGURE 1.24
A wall of notes, ideas and annotations (Chan, n.d.).

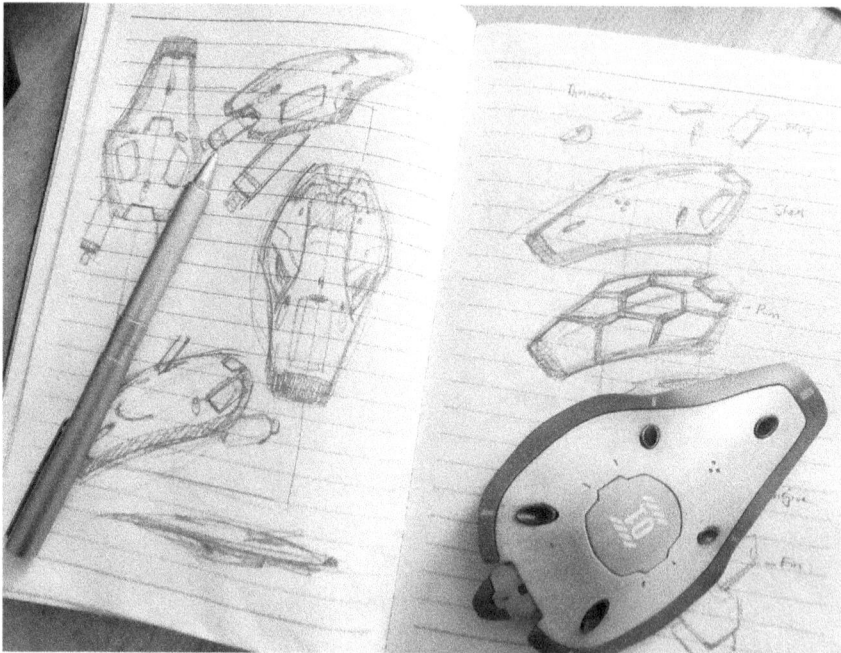

FIGURE 1.25
A Study Sketch and a hand-drawn General Arrangement Drawing showing an exploded view (Li, n.d.).

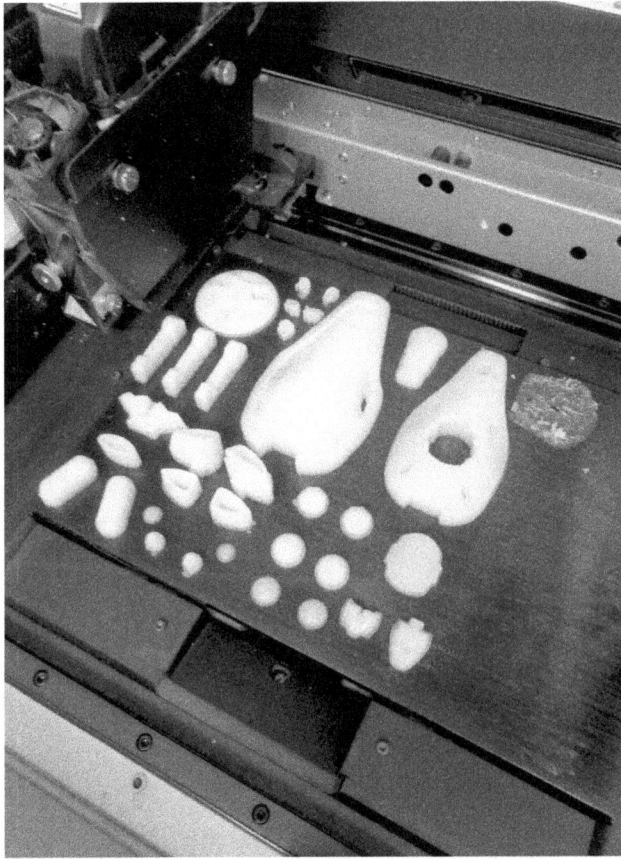

FIGURE 1.26
3D Printed parts using Material Jetting (Li, n.d.).

1.4 Stages of new product development

Based on the review of existing Models of new product development and the industrial designer's responsibilities, a simplified model of the industrial design process is presented as three stages: *concept, concept development* and *detail design*. As design activity progresses towards final specification of   design intent before manufacture, before manufacture, design development becomes increasingly detailed and specific. However, within this process, design will often iterate between the different stages of concept, development and detail design. As design progresses, the practitioner will continue to refine design intentions towards a single solution. This process of refinement is indicated by the reduction of choices as the design activity moves throughout the stages as shown in Figure 1.27 from Ulrich and Eppinger (2003).

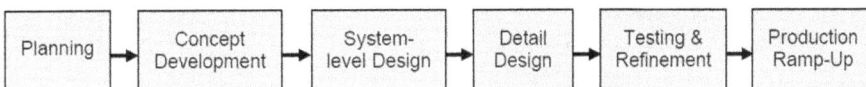

FIGURE 1.27
Ulrich and Eppinger's (2003) design model.

Throughout this process, various design tools are employed to embody and communicate design intent. The output of the *Concept Design* stage is an approximate description
of the form, function, and features of the artefact (Ulrich and Eppinger 2003). Sketching and
modelling are tools often used for the embodiment of concept ideas (Cross 2007; Ulrich and
Eppinger 2003). Design solutions are considered by all stakeholders in a process of

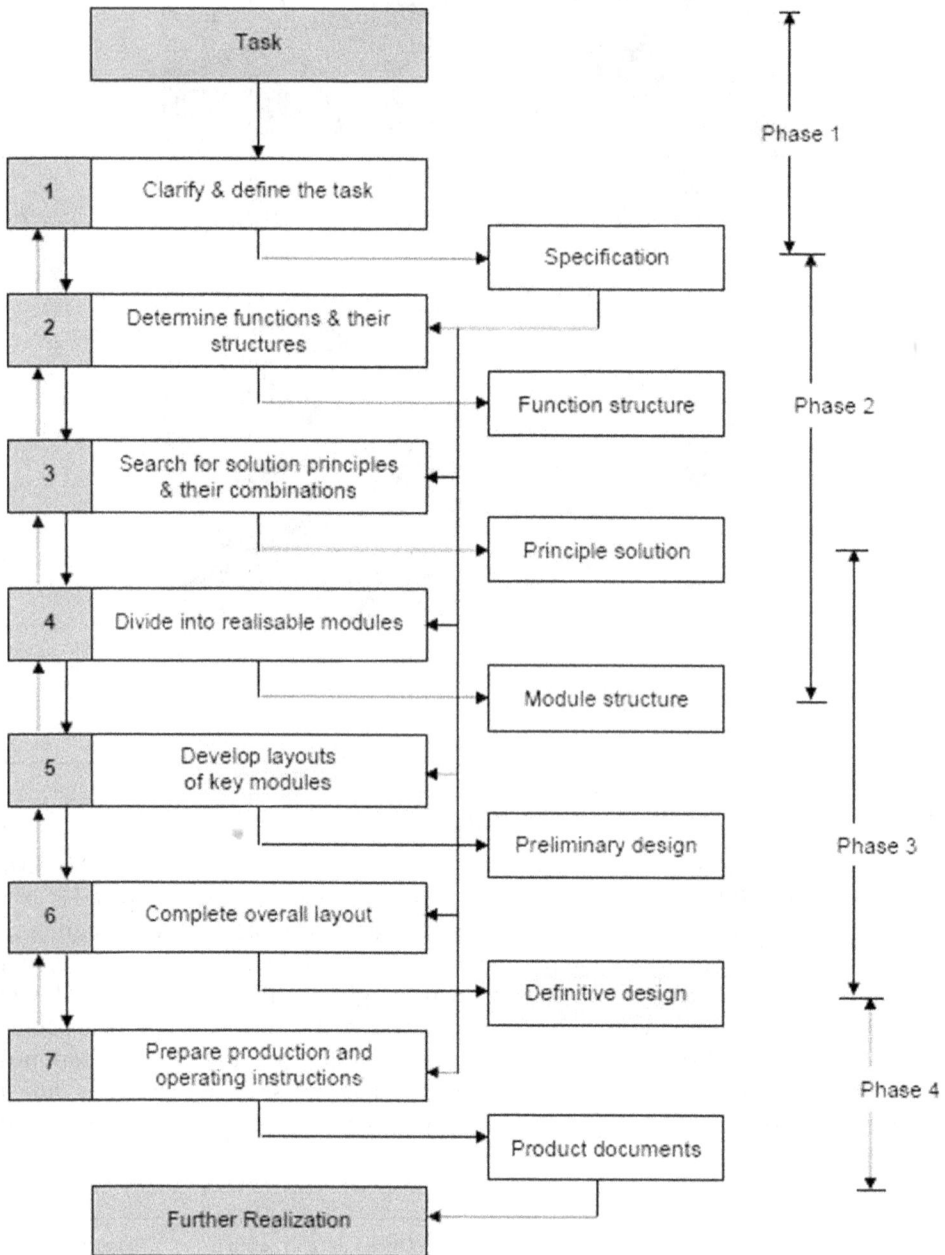

FIGURE 1.28
General approach to design according to VDI 2221.

selection, after which one or more concepts will be taken forward to be developed and refined. As with concept development and detail design, *concept design* requires an open, divergent approach to effectively explore the design problem through solution ideas and consideration for detail. This is represented by the VDI 2221 design model shown in Figure 1.28, and Figure 1.29 that shows the divergent and convergent processes involved.

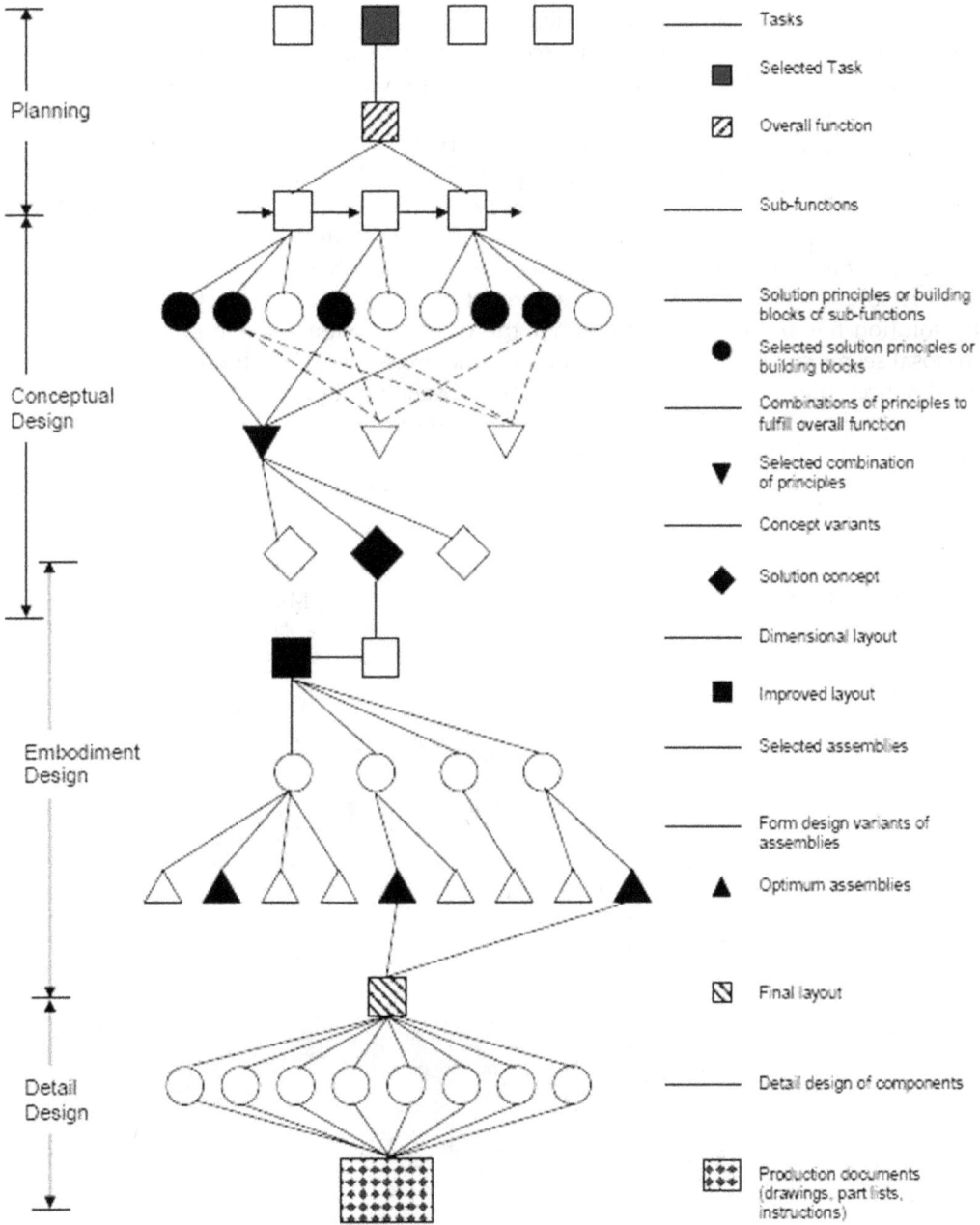

FIGURE 1.29
Divergence and convergence in the design process of VDI 2221.

VDI 2221 was developed by the German national standards body (DIN) as a systematic approach forthe development and design of technical systems and products.

A characteristic of concept development is its structured approach where 'communication of design information becomes increasingly important as the design becomes more detailed' (op cit., p. 130). Press 2003 describes the use of digital Models to communicate design intentions in parallel with the evolution of other design details. This more constrained phase of practice requires design solutions to be embodied in specific detail as the representations become increasingly focused on the manufactured outcome, 'The product's component parts and sometimes even the tooling required for manufacture must be depicted accurately and unambiguously' (Pipes 2007, p. 157). Next, *detail design* involves the movement of solution ideas from an embodiment of detail towards the exacting communication and specification of parts for final testing and manufacture. The search for solutions moves from the conceptual front end of practice, to converge towards a detailed solution proposal (Cross 2008, p. 195). *Detail design* is the final phase that aims to propose a specific solution to an often 'ill-defined design problem' (Cross 2007, pp. 23–25). Because of the ill-defined nature of the problem and the almost limitless possible solution proposals that may emerge, the design of the artifact ends not because the solution has been found, but rather the chance of significant improvement to the proposal seems small given the constraints of the project or due to limited time, resources and finances (Lawson 2004). Although several scholars defined the design process as different stages (Figure 1.30), the stages of new product development, together with their relation to the kinds of representations used, can be generally summarised within the three discrete steps which will be discussed in the next section.

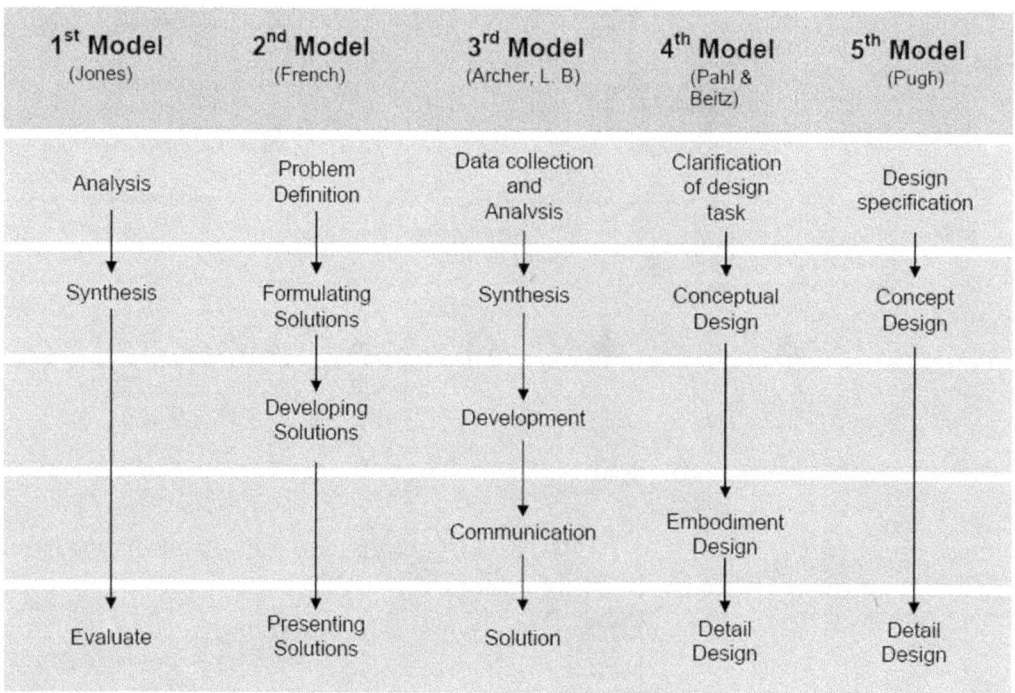

1st Model (Jones)	2nd Model (French)	3rd Model (Archer, L. B)	4th Model (Pahl & Beitz)	5th Model (Pugh)
Analysis	Problem Definition	Data collection and Analysis	Clarification of design task	Design specification
Synthesis	Formulating Solutions	Synthesis	Conceptual Design	Concept Design
	Developing Solutions	Development		
		Communication	Embodiment Design	
Evaluate	Presenting Solutions	Solution	Detail Design	Detail Design

FIGURE 1.30
Models of the design process.

1.4.1 *Concept design*

In the first phase of new product development, the *concept design* stage is mainly associated with idea generation activities even though the outlying problems may still be unclear. A large portion of this stage involves clarifying ideas through searching, establishing and selecting suitable concepts against known technical and economic specifications (French 1985; Pahl and Beitz 1996). This phase brings industrial design, engineering design and marketing teams together for the first time to make important early decisions. Once the function structures and system architectures are finalised, the physical design then takes place (Rosenthal 1992). This involves exploring design solutions usually with the use of pencil and paper to represent quick, spontaneous conceptual thoughts (Lawson 1984; Roozenburg and Cross 1991). *Concept design* is the first phase of new product development that involves generating ideas based on form, function, features, specifications and benchmarking with economic justification. Thinking, during concept design, is exploration focused. Design representation supports explorative design thinking through quick articulation of solution possibilities, often using Concept Sketches. Communication through representation is important as ideas are explored and discussed in terms of multiple stakeholders with different backgrounds, expertise, knowledge and criteria for evaluating the potential solutions (Figures 1.31–1.33). Examples of a visual Render and other studies are shown in Figures 1.34–1.36.

FIGURE 1.31
A user mapping exercise showing Scenarios and Storyboards (Chan, n.d.).

FIGURE 1.32
A typical focus group session with a facilitator (Chan, n.d.).

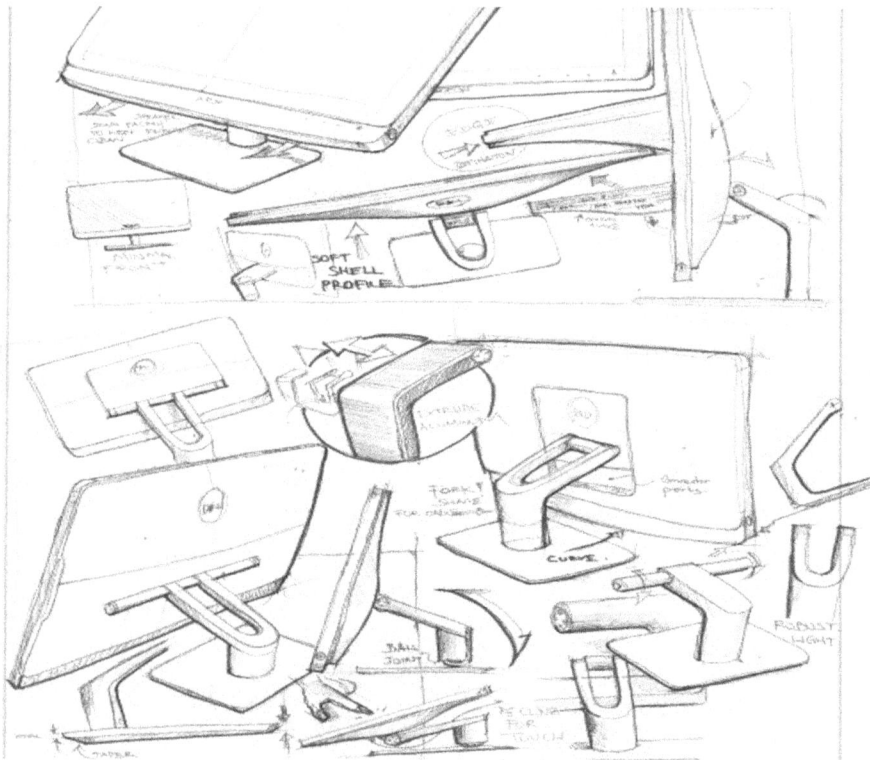

FIGURE 1.33
Examples of Study Sketches to explore the mechanism of a product (Chan, n.d.).

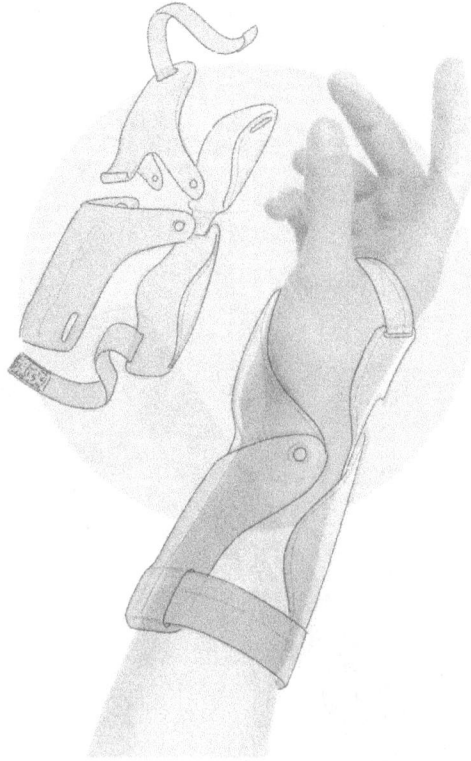

FIGURE 1.34
A Rendering of a wrist brace (Walters, n.d.).

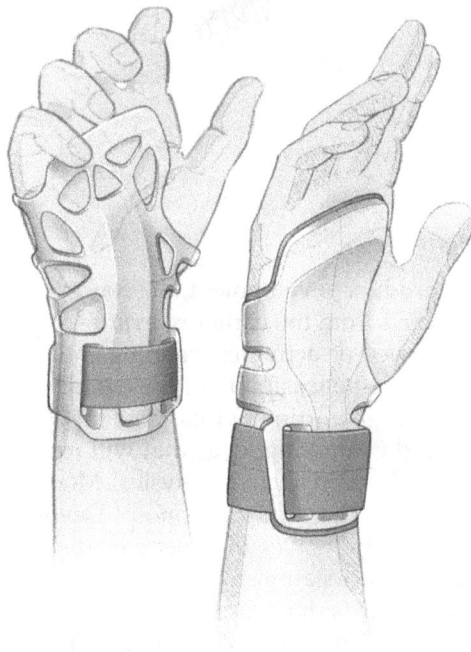

FIGURE 1.35
Another example of a Rendering (Walters, n.d.).

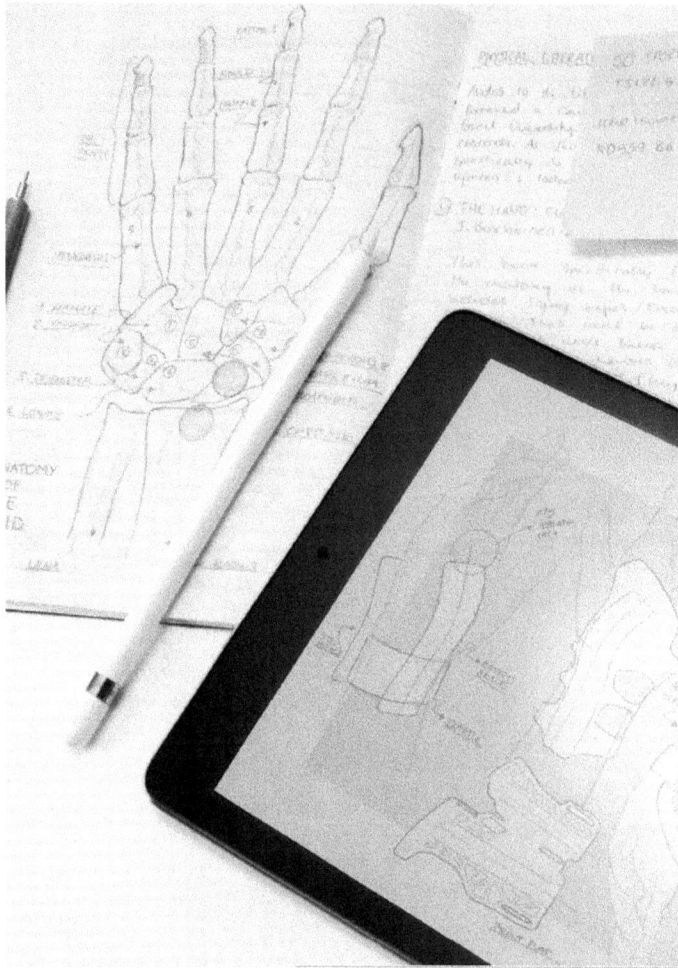

FIGURE 1.36
Study Sketches and ergonomic research (Walters, n.d.).

1.4.2 *Concept development*

In the second phase of new product development, the concept development stage follows the ideas that have been selected from the earlier *concept design* stage. This stage develops the initial ideas through a series of activities, refining them through extensive use of Sketches and Models to establish the feasibility of the overall concept (Cooper et al. 2000; Ulrich and Eppinger 2003). A large portion of this stage involves a higher detail of visual representation, developing and evaluating ideas that will meet the design specifications. Other representations frequently used include physical Models and Prototypes that define the arrangement and shape of the product. Concept Development is the second stage of new product development that involves the selection, development and evaluation of suitable concepts based on set specifications. The aims of *concept development* are to pivot away from exploration and towards development. With this change, communication of design intent increases, with a closer attention to detailed evaluation of the design (Figure 1.37–1.39), often through concept Models (Figure 1.40 and 1.41).

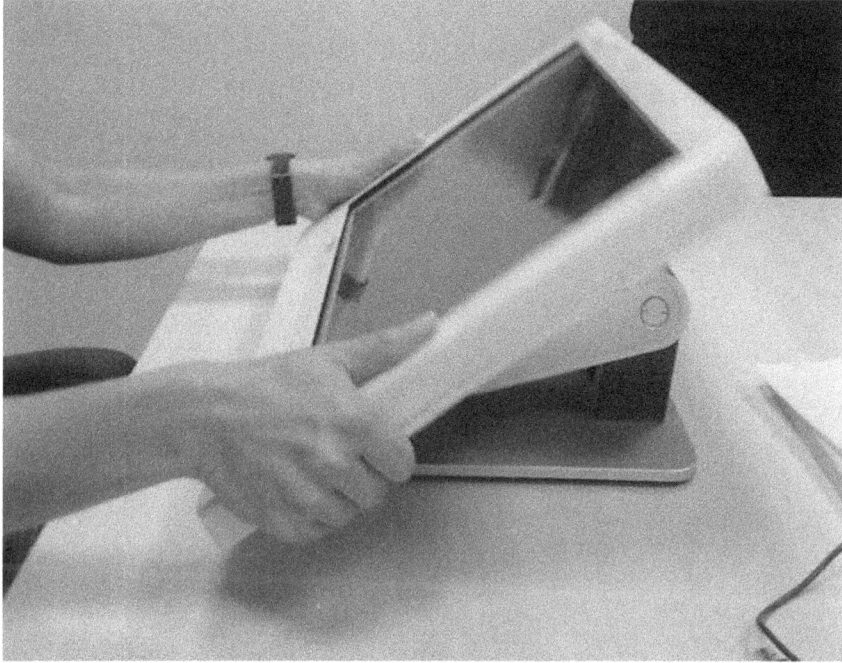

FIGURE 1.37
Example of an Experimental Prototype (Chan, n.d.).

FIGURE 1.38
CAD modelling showing internal components (Chan, n.d.).

FIGURE 1.39
An example of a Rendering (Chan, n.d.).

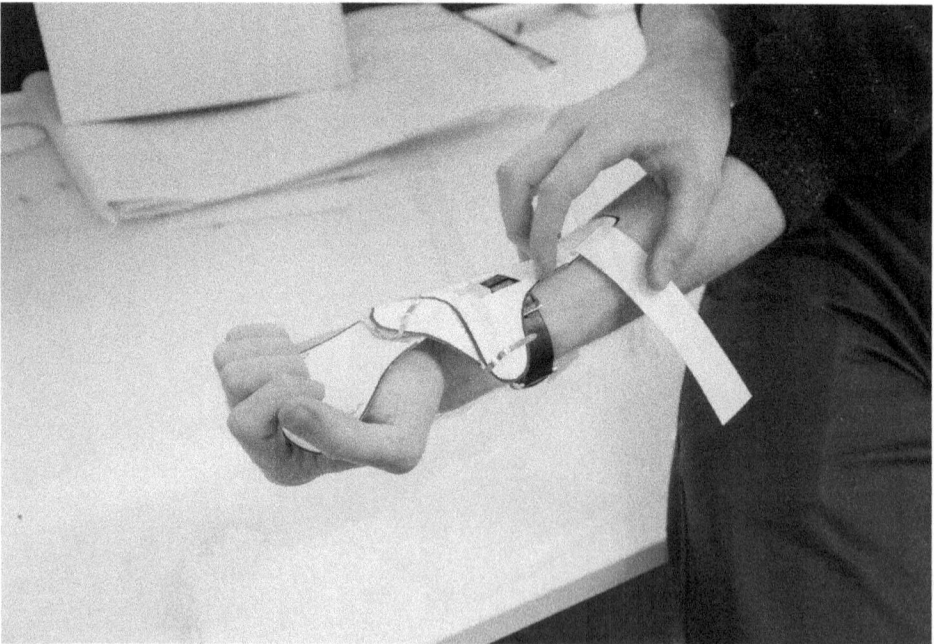

FIGURE 1.40
A Functional Concept Model in use (Walters, n.d.).

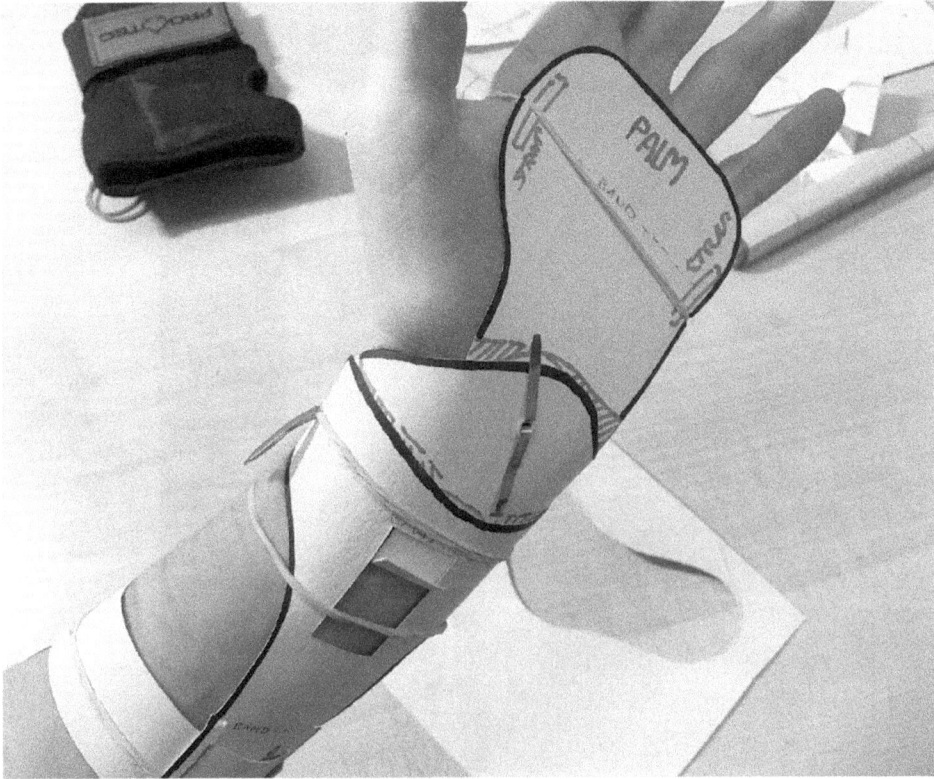

FIGURE 1.41
Another image of a Functional Concept Model in use (Walters, n.d.).

1.4.3 *Detail design*

In the third stage of new product development, the *detail design* phase is concerned with many small but important aspects of the product prior to manufacture (Haskell 2004). This phase produces a final and highly detailed technical description of each component including the materials, colours, finishing, surface texture properties, fits, tolerances, positioning and assembly details. Outputs at this stage may include technical descriptions, such as General Arrangement Drawings, that incorporate both layout design (arrangement of components) and form (aesthetics) in consideration to technical and economic constraints (Dym and Little 2009). Other activities include checking and testing of components and product features prior to manufacture (Pahl and Beitz 1996). Cooper (2018) claimed that as construction (or production) nears, precision becomes more important, and any imprecision must be eliminated. Detail design is the final stage of new product development that realises the physical product through the specification of details such as material, size, assembly, etc., with final testing before the actual mass production. *Detail design* is concerned with an unambiguous definition of the design intent. Rather than providing an opportunity for interpretation, design representations at this stage are employed to answer specific technical, operational, manufacturing and aesthetic questions (Figures 1.42–1.46). As thinking turns towards the detailed specification, design representations and the thinking associated with it pivot towards certainty, identification and resolution of design problems at a micro level.

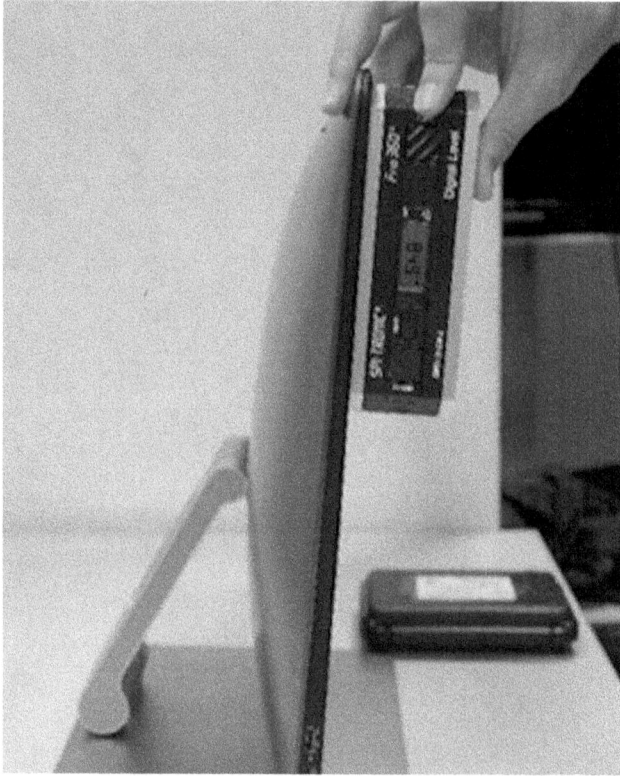

FIGURE 1.42
A Pre-Production Prototype (Chan, n.d.).

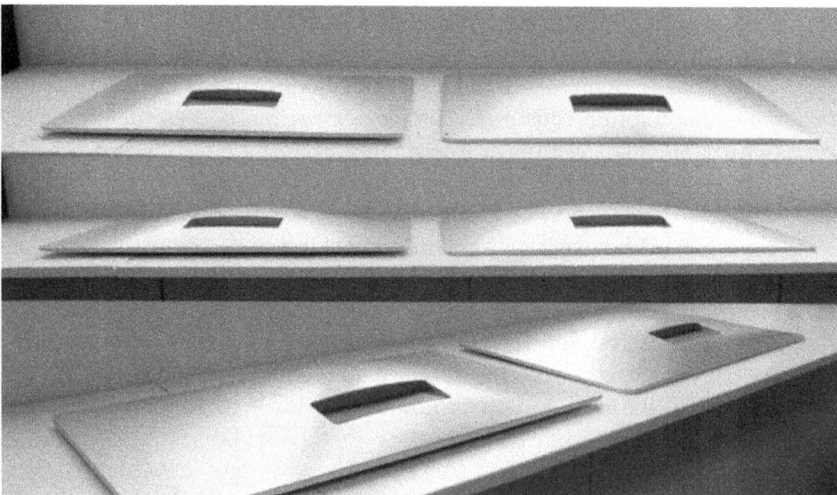

FIGURE 1.43
Off-Tool Prototypes being produced (Chan, n.d.).

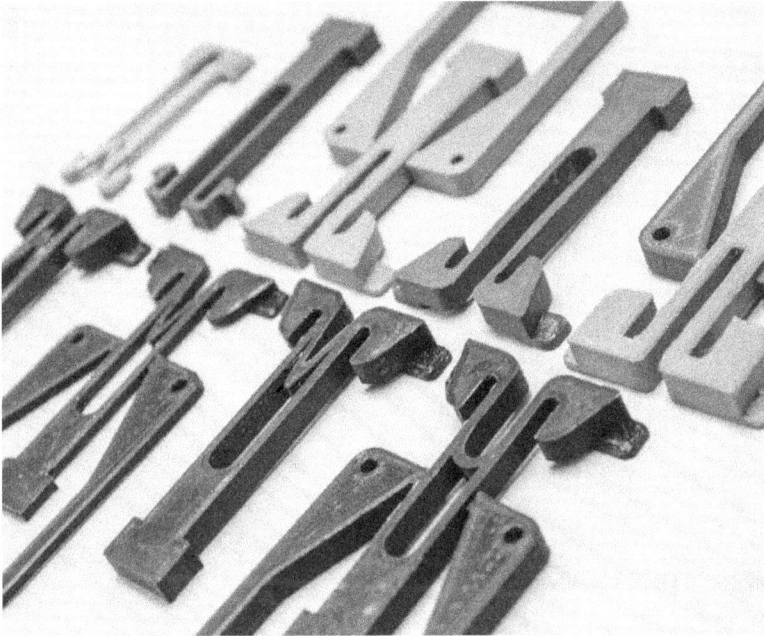

FIGURE 1.44
Experimental Prototypes (Walters, n.d.).

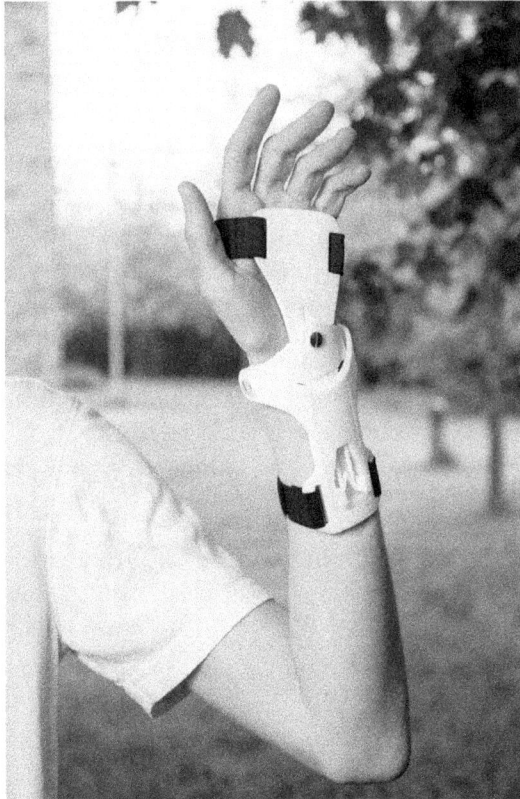

FIGURE 1.45
A Concept of Operation Model in use (Walters, n.d.).

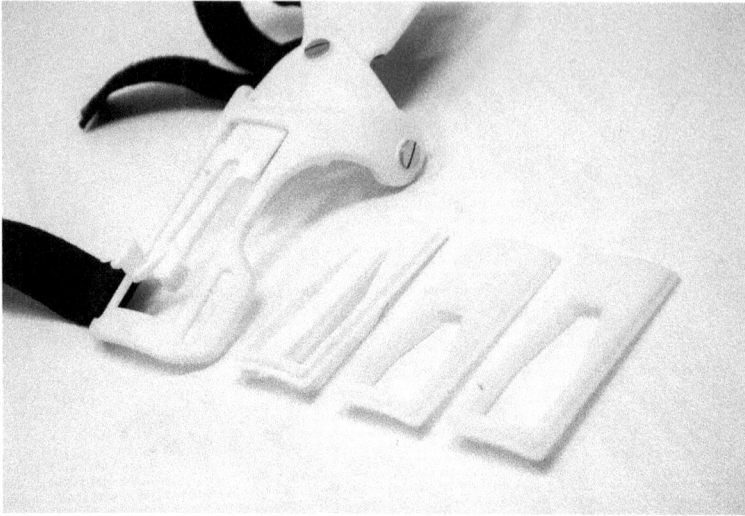

FIGURE 1.46
Parts of the Concept of Operation Model with test pieces (Walters, n.d.).

1.5 Summary

In this chapter, we have outlined the relationship between design and design re-presentation. We have discussed design practice in relation to the nature of design pro-blems, and how designers must express potential solution candidates through design representation. This has included touching on how representation may relate to design thinking. A theme that will be further expanded on in Chapter 2. Industrial design and the industrial designer's role and responsibilities were outlined to position the book's focus on a discipline involved with the design and production of physical products.

The historic foundations of industrial design were outlined in summary as emerging from the industrial revolution due to the necessity to specify design intent prior to production. Industrial design's dual focus on both technical functions, through the ap-plication of science and engineering, together with aesthetics derived from the arts, has been discussed. The work and responsibilities of the industrial designer were further discussed in relation to new product development, and a requirement to communicate design intent across various stakeholders (Section 1.4).

The chapter has also described a process of industrial design as moving through three discrete stages: *concept, concept development* and *detail design*. The types of works, re-sponsibilities and designs were discussed with a focus on how different representations are required to satisfy the changing requirements of an industrial design process that moves through these three stages. The critical relationships between representational types and phases of the design process will be further expanded on in Chapter 2.

References

Andreasen, M. M., Kähler, S., et al. (1988) *Design for Assembly* (2nd ed.). London: IFS Publishers. Quoted in: "A Typology of Designs and Designing," *International Conference on Engineering Design ICED 03*, Stockholm.

Archer, L. B. (1965) *Systematic Method for Designers*. London: The Design Council. Quoted in: Jones, J. C. (1970) *Design Methods* (2nd ed.). New York: John Wiley and Sons, Inc.

Ashford, F. C. (1969) *The Aesthetics of Engineering Design*. Quoted in: Hague, R., Campbell, I., Dickens, P. and Reeves, P. (2002) Integration of Rapid Manufacturing in Design. *Time Compression Technologies*, Vol. 10(4), p. 48.

Billings, S. (2006) Functional Fashion. *Design Week*, Vol. 21(37).

Bohemia, E. (2002) Designer as Integrator: Reality or Rhetoric? *Design Journal*, Vol. 5(2).

Bousbaci, R. (2008) Models of Man. In Design Thinking: The "Bounded Rationality "Episode." *Design Issues*, Vol. 24(4).

Bradshaw, J. L. and Nettleton, N. C. (1983) *Human Cerebral Asymmetry*. Prentice Hall. Quoted in: Tovey, M. (1984) Designing with Both Halves of the Brain. *Design Studies*, Vol. 5(4).

Bucciarelli, L. (1994) *Designing Engineers*. Cambridge, MA: MIT Press.

Bürdek, E. E. (2005) *Design - History, Theory and Practice of Product Design*. Basel, Switzerland: Birkhauser.

Burghardt, M. D. (1999) *Introduction to Engineering Design and Problem Solving*. Singapore: McGraw-Hill International.

Campbell, R. I., Hague, R. J., et al. (2006) The Potential for the Bespoke Industrial Designer. *The Design Journal*, Vol. 6(3), pp. 24–26.

Cooper, D. (2018) Imagination's Hand: The Role of Gesture in Design Drawing. *Design Studies*, Vol. 54, pp. 120–139.

Cooper, R., Bruce, M., et al. (2000) The Designer as Innovator. Quoted in: Scrivener, S. A. R., Ball, L. J. and Woodcock, A. (2000) Collaborative Design. *Proceedings of CoDesigning 2000*. London: Springer-Verlag, p. 117.

Corfield, K. G. (1979) *Product Design: A Report by Mr K G Corfield Carried Out for the National Economic Development Council*. London: National Economic Development Council.

Cross, N. (1983) Methods Guide: Control of Technology. *3rd Level Course Unit 8 T361*. Milton Keynes: Open University Press.

Cross, N. (1984) *Developments in Design Methodology*. Chichester: John Wiley and Sons Ltd.

Cross, N. (2007) *Designerly Ways of Knowing*. Basel, Switzerland: Birkhauser Press.

Cross, N. (2008) *Engineering Design Methods: Strategies for Product Design* (4th ed.). Chichester: John Wiley and Sons.

Cross, N. (2021) *Engineering Design Methods: Strategies for Product Design* (5th ed.). Chichester: John Wiley and Sons Ltd.

Dictionary of Art Terms (2003) *Thames and Hudson World of Art*. E. Lucie-Smith (Ed.). London: Thames and Hudson.

Dorta, T. (2005) Hybrid Modeling: Manual and Digital Media in the First Steps of the Design Process. http://www.din.umontreal.ca/documents/dorta/9-eCAADe2005.pdf. Accessed on 12 October 2008.

Dorta, T., Perez, E., et al. (2008) The Ideation Gap: Hybrid Tools, Design Flow and Practice. *Design Studies*, Vol. 29(2), pp. 121–141.

Dym, C. L. and Little, P. (2009) *Engineering Design: A Project-Based Introduction* (2nd ed.). New Jersey: John Wiley and Sons Inc.

Eissen, K. and Steur, R. (2008) *Sketching: Drawing Techniques for Product Designers*. Singapore: Bis Publishers/Page One Publishing.

Erlhoff, M. (1987) Kopfüber zu Füßen. Prolog für Animateure. In: documenta 8, Vol. 1 Kassel. Quoted in: Bürdek, Ernhard E. (2005) *Design: History, Theory and Practice of Product Design*. Basel: Birkhauser.

Evans, M. and Wormald, P. (1993) The Future Role of Virtual and Physical Modelling in Industrial Design. *IDATER 1993 Conference*. Loughborough University (Design and Technology).

Fiell, C. and Fiell, P. (2003a) *Design for the 21st Century*. Koln: Taschen GmbH.

Fiell, C. and Fiell, P. (2003b) *Industrial Design A-Z*. Cologne: Taschen GmbH.

French, M. J. (1985) *Conceptual Design for Engineers*. Berlin: Springer Verlag.

Frishberg, N. (2006) *Prototyping with Junk. Interactions* (January/February).

Garner, S. (1999) *Drawing and Designing: An Analysis of Sketching and Its Outputs as Displayed by Individuals and Pairs When Engaged in Design Tasks* (unpublished PhD thesis). Loughborough: Loughborough University.

Garner, S. (2006) *Modelling Workbook 1: T211 Design and Designing Workbook 1 Technology Level 2* (2nd ed.). Milton Keynes: The Open University Press.

Gemsera, G. and Leenders M. A. A. M. (2001) How Integrating Industrial Design in the Product Development Process Impacts on Company Performance. *Journal of Product Innovation Management*, Vol. 18 (1), pp 28–38.

Goel, V. (1995) *Sketches of Thought*. Cambridge MA: MIT Press.

Goldschmidt, G. (1994) On Visual Design Thinking: The Vis Kids of Architecture. *Design Studies*, 15(2), pp. 158–174.

Goldschmidt, G. (1995) The Designer as a Team of One. *Design Studies*, Vol. 16(2), pp. 189–209.

Hack, G. and M. Canto (1984) Collaboration and Context in Urban Design. *Design Studies*, Vol. 5(3), pp. 178–184.

Hannah, B. (2004) *Becoming a Product Designer*. New Jersey: John Wiley and Sons.

Haskell, B. (2004) *Portable Electronics: Product Design and Development*. New York: McGraw Hill.

Heskett, J. (1980) *Industrial Design*. London: Thames and Hudson Ltd.

Hirdina, H. (1998) Voraussetzungen postmodern Designs. In: Flierl, B. and Hirdona, H. (1995) *Postmoderne und Funktionalismusm*. Berlin: Sechs Vorträge. Quoted in: Bürdek, E. E. (2005) *Design: History, Theory and Practice of Product Design*. Basel: Birkhauser.

Holme, G. (1934) *Industrial Design and the Future: A Challenge to the Producer*. London: The Studio Limited.

IDSA (2006) *Industrial Designers Society of America (IDSA): About Industrial Design*. http://www.idsa.org/webmodules/articles/anmviewer.asp?a=89andz=23. Accessed on 2 February 2006.

Jones, C. J. (1969) *Design Methods in Architecture*. London. Quoted in: Bürdek, E. E. (2005) *Design: History, Theory and Practice of Product Design*. Basel: Birkhauser.

Jones, C. J. (1992) *Design Methods* (2nd Ed.). New York: John Wiley and Sons.

Kavakli, M., Stephen A. R., et al. (1998) Structure in Idea Sketching Behaviour. *Design Studies*, Vol. 19, pp. 485–517.

Kosslyn, S. (1994) *Image and Brain: The Resolution of the Imagery Debate*. Cambridge, MA: MIT Press.

Kristensen, T. (1995) A Contribution of Design to Business: A Competence-Based Perspective. *Design Management Proceedings of the European Academy of Design*, 11–13 April 1995, University of Salford.

Lawson, B. (1984) Cognitive Strategies in Architectural Design. In: Cross, N. (ed.) (1984).

Lawson, B. (2004) *What Designers Know*. Oxford: Architectural Press.

Lawson, B. (2006) *How Designers Think: The Design. Process Demystified* (4th ed.). Oxford: Oxford University Press.

McKim, R. H. (1980) *Experiences in Visual Thinking*. Boston: PWS Publishers.

Pahl, G. and Beitz, W. (1996) *Engineering Design: A Systematic Approach*. (2nd ed.). New York: Springer-Verlag.

Persson, S. (2002) *Industrial Design: Engineering Design Interaction: Studies of Influencing Factors in Swedish Product Developing Industry*. Thesis for the Degree of Licentiate of Engineering, Product and Production Department. Göteborg, Sweden: Chalmers University of Technology.

Persson, S. (2005) *Toward Enhanced Interaction between Engineering Design and Industrial Design*. PhD Thesis. Goteborg, Sweden: Chalmers University of Technology.

Pipes, A. (1990) *Drawing for 3-Dimensional Design: Concepts, Illustration, Presentation*. London: Thames and Hudson.

Pipes, A. (2007) Drawing for Designers. London: Laurence King Publishing. Quoted In: Purcell, A. T. and Gero, J. S. (1998) Drawings and the Design Process: A Review of Protocol Studies in Design and Other Disciplines and Related Research in Cognitive Psychology. *Design Studies*, Vol. 19(4), pp. 389–430.

Press, M. and Cooper, R. (2003) *The Design Experience: The Role of Design and Designers in the Twenty-First Century*. London: Routledge.

Pugh, S. (1991) *Total Design: Integrated Methods for Successful Product Engineering*. Essex, UK: Pearson Education Limited/Addison-Wesley Publishers Ltd.

Purcell, A. T. and Gero, J. S. (1998) Drawings and the Design Process: A Review of Protocol Studies in Design and Other Disciplines and Related Research in Cognitive Psychology. *Design Studies*, Vol. 19(4), pp. 389–430.

Rittel, H. and Webber, M. M. (1973) Dilemmas in a General Theory of Planning. *Policy Sciences*, Vol. 4, pp. 155–169.

Robertson, T. (1996) Embodied Actions in Time and Place: The Cooperative Design of a Multimedia Educational Computer Game. *CSCW*, Vol. 5(4), pp. 341–367. Quoted in: Perry, M., and Sanderson, D. (1998) Coordinating Joint Design Work: The Role of Communication and Artifacts. *Design Studies*, Vol. 19(3), pp. 273–288.

Rodriguez, W. (1992) *The Modelling of Design Ideas: Graphics and Visualization Techniques for Engineers*. Singapore: McGraw-Hill Book Company.

Roozenburg, N. and Cross, N. (1991) Models of the Design Process: Integrating across the Disciplines. *Design Studies*, Vol. 12(4), pp. 215–220.

Roozenburg, N. and Eekels, J. (1995) *Product Design: Fundamentals and Methods* (2nd ed.). Chichester: John Wiley and Sons Ltd.

Rosenthal, S. R. (1992) *Effective Product Design and Development: How to Cut Lead Time and iNcrease Customer Satisfaction*. Homewood, IL: Business One Irwin.

Schön, D. A. (1983) *The Reflective Practitioner: How Professionals Think in Action*. London: Ashgate.

Sebastian, R. (2005) The Interface between Design and Management. *Design Issues*, Vol. 21(1), pp. 81–93.

Shetty, D. (2002) *Design for Product Success*. Dearborn, MI: Society of Manufacturing Engineers (SME).

Sparke, P. (1983) *Consultant Design: The History and Practice of the Designer in History*. Pembridge History of Design Series (3rd ed.). London: Pembridge Press.

Stempfle, J. and Badke-Schaub, P. (2002) Thinking in Design Teams: An Analysis of Team Communication. *Design Studies*, Vol. 23(5), pp. 473–496.

Temple, S. (1994) Thought Made Visible: The Value of Sketching. *Co-Design*, Vol. 1, pp. 16–25.

Tjalve, E. (1979) *A Short Course in Industrial Design*. London: Butterworth and Co.

Tjalve, E., Andreasen, M. M., et al. (1979) *Engineering Graphic Modelling: A Workbook for Design Engineers*. London: Butterworth and Co.

Tovey, M. (1984) Designing with Both Halves of the Brain. *Design Studies*, Vol. 5(4), pp. 219–228.

Tovey, M. (1989) Drawing and CAD in Industrial Design. *Design Studies*, Vol. 10(1), pp. 24–39.

Tovey, M. (1997) Styling and Design: Intuition and Analysis in Industrial Design. *Design Studies*, Vol. 18(1), pp. 5–31.

Ulrich, K. T. and Eppinger, S. D. (2003) *Product Design and Development* (3rd ed.). New York: McGraw-Hill.

Urban, G. L. and Hauser J. R. (1993) *Design and Marketing of New Products* (2nd ed.). Englewood Cliffs, NJ: Prentice Hall. Quoted in: Song, X., Michael, M., Mitzi, M. and Schmidt, J. B. (1997) Antecedents and Consequences of Cross-Functional Cooperation: A Comparison of R&D, Manufacturing and Marketing Perspectives. *Journal of Product Innovation Management*, No. 14.

Utterback, J. B. and Vedin, A. (2006) *Design Inspired Innovation*. Singapore: World Scientific Publishing.

Verstijnen, I. M., Hennessey, J. M. et al. (1998) Sketching and Creative Discovery. *Design Studies*, Vol. 19(4), pp. 519–546.

Vissler, W. (2006a) *The Cognitive Artifacts of Designing*. New York: Routledge.

Vissler, W. (2006b) Designing as Construction of Representations: A Dynamic Viewpoint in Cognitive Design Research. *Human-Computer Interaction,* Special Issue, Foundations of Design in HCI, Vol. 21(1), pp. 103–152.

Wang, L., Shen, W., et al. (2002) Collaborative Conceptual Design: State of the Art and Future Trends. *Computer Aided Design* 34(13), pp. 981–996.

Woodham, J. M. (1983) *Pembridge History of Design Series: The Industrial Designer and the Public.* London: Pembridge Press.

Wright, I. (1998) *Design Methods in Engineering and Product Design.* Berkshire: McGraw Hill Publishing Company.

Yamamoto, M. and Lambert, D. R. (1994) The Impact of Product Aesthetics on the Evaluation of Industrial Products. *Journal of Product Innovation Management,* 11(4), pp. 309–324.

2

Design thinking through representation

2.1 Design representation

There are several and often conflicting definitions of design to be found in the literature (Cross 2011). However, two concepts are of particular for our current thesis for re-presentation's importance to design. First, design is a futurist activity. Design is concerned with the articulation, expression and communication of possible, but yet-to-be future ser-vices, products and experiences (Nelson and Stolterman 2003, p. 135). And, since the practice of design is often separate from the implementation of a design through pro-duction, manufacture, construction, and so on, design's purpose is to specify intent towards possible solution candidates (Cross 2007, p. 115).

In order to specify design intent, designers use a variety of tools to represent their design ideas as part of a process of design conceptualisation and development. These expressions of intent, afforded through and influenced by the media of expression, support the exploration, development and detailed specification of solution ideas. The necessity to represent intent towards solution candidates can be traced back to design's concern with the solution of ill-defined problems (Rittel and Webber 1973). And, as ill-defined problems are often associated with issues related to the humanities, in contrast with problems of the natural sciences, they can be further described as involving com-plex, adaptive systems with multiple stakeholders, interconnections, variables or possi-bilities for change, adaption or re-orientation. The type of problems that design attempts to address are often culturally sensitive, moving dynamically in line with broader shifts in societal norms, conventions, values, morals and traditions. Thus, design representation provides opportunity to engage the complexity of design problems through the ability of representation to identify a variety of possible solution candidates.

In an early study, Rittel and Webber (1973) described the complexity of ill-defined design problems within the context of design. Because of the ill-defined nature of design problems, the act of design seeks to arrive at the most appropriate solution, rather than to seek an objectively true or correct answer; the norm for scientific discovery. To arrive at a most ap-propriate solution, design is required to speculate on future scenarios to offer improvements to an existing state. The necessity to speculate on propositional solutions, and to assess their potential to offer improvement, requires the generation of design representations as approx-imations of solution ideas (Cross 2008; Visser 2006; Goldschmidt 1997). These representations then act as a means through which a potential design is identified, evaluated and developed. In the expression of intent through design representation, the designer adopts a solution focus. Through representation, the designer evaluates the potential of solution ideas as approxima-tions of solutions to ill-defined design problems. Rather than only relying on internal re-presentations within the mind, as is the case with simpler tasks that require the application of

DOI: 10.1201/9781003227694-2

methods, skills and/or knowledge to reach a known goal, design necessitates external re-presentation as means to distribute cognitive load, thereby enabling better navigation towards a yet unknown solution. This distribution of cognition has also been described as a reflection on representation (Schön 1983). These reflections interact with internal representations in the mind, so as to inform an understanding of both the nature of the design problem and possible solutions. Figure 2.1 shows several Idea Sketches produced by Natasha Dareshani in which the use of orange-coloured lines represents further lines of enquiry during the design process. She also uses symbols to represent information within the context of the idea.

FIGURE 2.1
Idea Sketches showing different concepts (Dareshani, n.d.).

A constructed and distributed approach to understand design representation also re-lates back to the nature of design problems. The act of representation has been equated with a need for design to deal with complex problems in contextual situations of un-certainty. As discussed above, design problems are complex and ill-defined in their concern for adaptive systems, involving multiple stakeholders that are also influenced by human action and emotion. Design problems can be characterised by their sensitivity for dealing with the full range of human needs, from the emotional to the functional. To frame these design problems in a more manageable way, the designer externalizes cog-nition through representation to express solution proposals in an effort to understand the nature, scope and detail of both problem and potential solutions (Cross 1992). This process has also been described as the identification of a problem-solution pair.

If in tackling ill-defined design problems the designer often employs the construction of representations to support interaction between internal mental images and external ex-pressions of design intent, the character of a design representation will influence the interpretation and evaluation of a potential design solution. After better understanding the requirements of the product being designed; its associated problems and opportu-nities for improvement, the designer will eventually arrive at a specific and detailed design proposal (Cross, op cit., p. 21). Through this process of solution proposition and problem development, a variety of design representations are employed that attempt to approximate, at various levels of detail and fidelity, form and functional aspects of an intended design. This is shown in Valentina Dermachi's work where she has produced several working Experimental Prototypes for a garden seed planter in Figures 2.2 and 2.3.

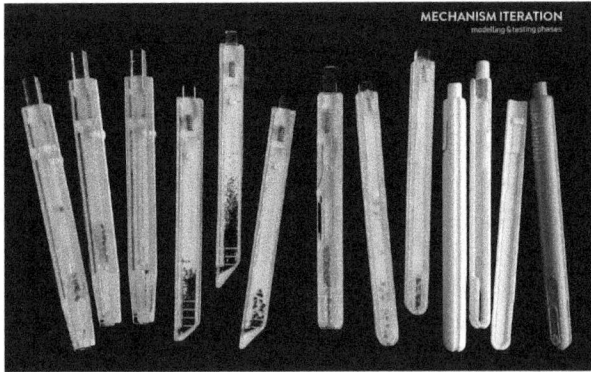

FIGURE 2.2
Experimental Prototypes (Dermachi, n.d.).

FIGURE 2.3
Experimental Prototypes in use (Dermachi, n.d.).

From a conceptual and explorative phase in design (i.e. *Concept Design*) through to the prescriptive requirements of *Detail Design* (Baxter 1995; Ulrich and Eppinger 2003; Visser 2006, p. 115), embodiments of intent, achieved through the use of Sketches, Drawings, Models and Prototypes, support the generation and evolution of design ideas throughout design process. If the representation of design solutions is critical to the means through which ill-defined design problems and their solutions are explored, identified and re-fined, the media of representation has implications for how design ideas are identified and communicated during the design process.

2.1.1 Design representation and media of expression

The digital revolution has ushered in advances in the types of design representations used in design practice during new product development. Traditionally, design intentions were represented, reflected on and developed through analog hand-drawing, sketching and craft-based model making methods and techniques. However, with the continuing impact of digital and hybrid tools and media of expression, the designer is presented with an increasing variety of media to employ in pursuit of expression of design intent through representation such as 3D CAD representations shown in Figures 2.4 and 2.5.

FIGURE 2.4
A Multi-View Drawings and a CAD Rendering (Aligizakis, n.d.).

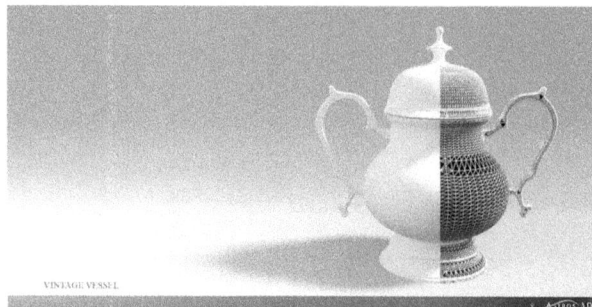

FIGURE 2.5
A CAD Rendering (Aligizakis, n.d.).

Although an increasing variety of tools bring benefits, limitations in the use of 3D CAD software have been identified (Bilda and Demirkan 2003; Goel 1995; Goldschmidt 1992; Dorta et al. 2008; Lawson 2004; Jonson 2002; Robertson and Radcliff 2009; Tovey and Owen 2000). Some scholars claim that the use of 3D CAD may restrain the exploration of design ideas, constraining the ways in which ideas are developed due to the particular affordances and limitations of digital software tools and their interface. While less of a problem for more experienced designers, a novice designer may find the use of 3D CAD can inhibit the flow of creative ideas. This may be due to the command-based structure of most CAD modelling tools, together with time investments required in the building of CAD Models, resulting in increased fixation and attachment to concept.

Even with the emergence of an increasing variety of other digital design tools such as rapid prototyping, additive manufacturing, graphics tablets, haptic technologies and virtual reality (Evans et al. 2004), sketching still continues to be valued as a core competency for Industrial Designers (Pipes 2007, p. 12). Powell (op cit.) described a designer's ability to represent intent through sketching as critical to having a conversation with themselves to keep pace with the speed of one's own thinking (ibid, p. 6). If this is the case, the type of media used in the representation and communication of design intent appears to play a role as potential influence on cognition during design and development, as evidenced by the impact of CAD tools. It may be that sketching, as media of expression, is particularly effective in providing opportunity for reflection between the external expression of ideas and the designer's own thinking towards possible design solutions, and their development.

2.1.2 Design representation as construction

Visser (2006, p. 119) provides the term *construction of representations* to describe a process-oriented view of design embodiment through representation. As part of this process, designers employ imprecise external representations of design intent, which then interact with internal mental images known in cognitive sciences as internal representations. This distributed interaction between external representation and mental image provides a framework for interpretation of design intent. Adopting Visser's (ibid) approach, interpretation is a synthesis of the characteristics of the representation itself and the subjective interpretation of its meaning. In this sense, design representations act as stimulation to reform or revise a metal image. Internal representations can be described as the constructed thoughts and imaginations of the designer, while external representations are produced as approximations of design intent. However, according to Schön's (1983) reflective-practice paradigm for understanding design activity, representation is closer to an act of distributed cognition, whereby thinking towards potential solutions happens as part of the act of representation. In this sense, representations extend cognition beyond the mind. Through the reflective-practice paradigm for understanding design thinking, representation is closer to the theory of embodied cognition as representations are employed as extensions beyond the brain extensions beyond the brain. For example, Figure 2.6 shows an early sketch by the designer, Chan, who designed the Glenfiddich Whiskey Glorifier (Figure 2.7) which was later realised using Computer Aided Design as a photo-realistic representation in Figure 2.8; and finally photographed as an actual product in Figure 2.9.

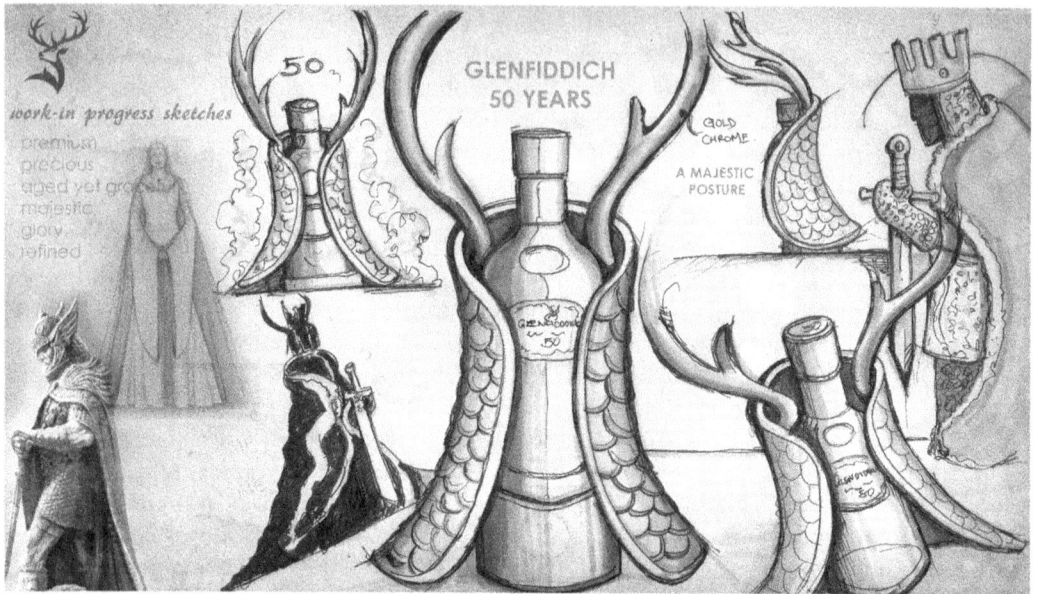

FIGURE 2.6
Referential Sketches (Chan, n.d.).

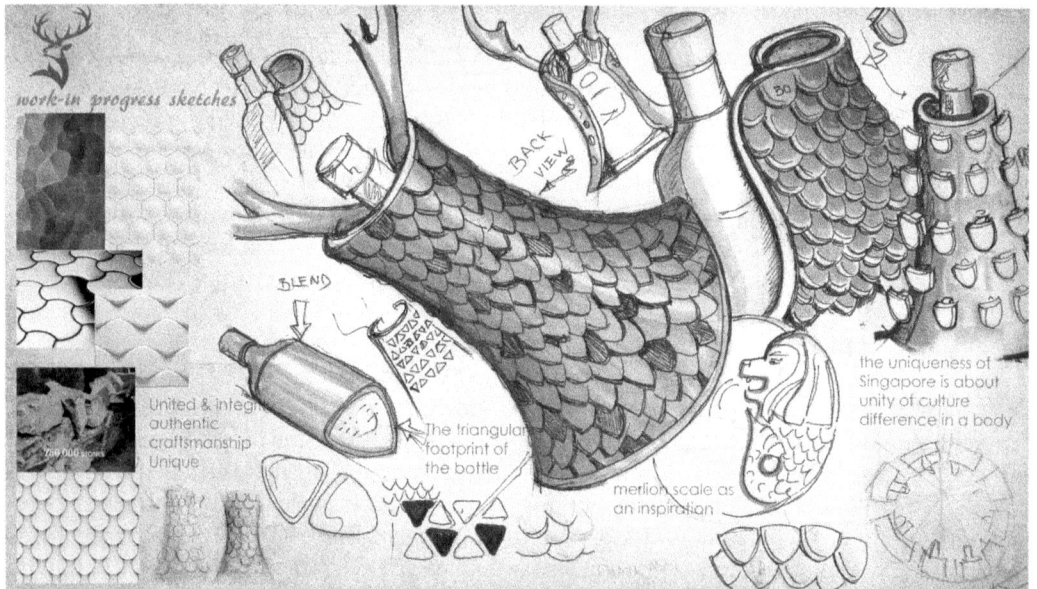

FIGURE 2.7
An example of an Inspiration Sketch (Chan, n.d.).

FIGURE 2.8
A Rendering of a product (Chan, n.d.).

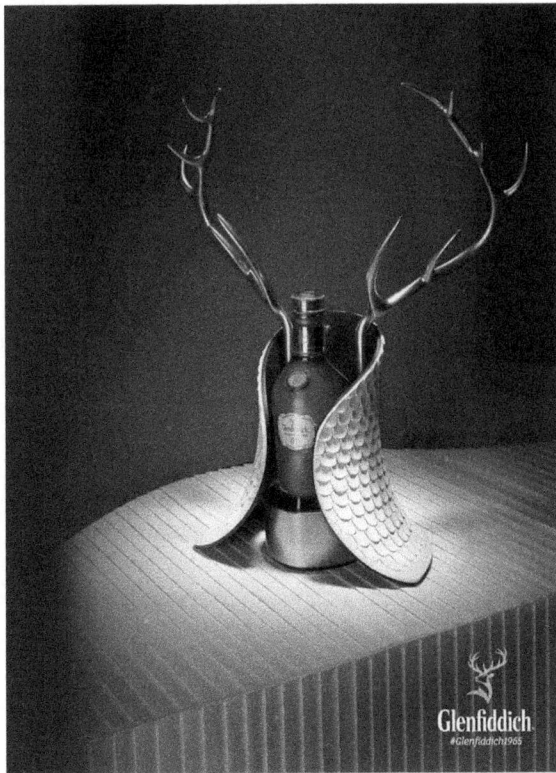

FIGURE 2.9
Actual product photograph.

2.2 Representation and design cognition

Adopting a distributed approach to understand the role and use of representation in design offers an opportunity to consider representation's role in design thinking from various positions. Unfortunately, there has been little work to date exploring relationships between design representation and cognition. The reflective-practice paradigm (Schön 1983) is perhaps one of the most influential theories of design activity. The following sections consider relations between design representation as distributed cognition and design activity as reflective-practice (Schön 1983). In particular, we examine representational ambiguity and fidelity as constructs to identify different types of design representation; their relation to a distributed approach to understanding the role and use of representation in design.

2.2.1 Design representation as reflective-practice

Design activity in the form of *reflective practice* has been acknowledged by design scholars including Cross (2007, 2008), Goel (1995), Jonson (2005), Stolterman et al. (2008), Lawson (2006), Schön (1983) and Tovey et al. (2003). The concept of *reflective practice* can broadly be defined as an approach to understanding design from a constructed point of view. Distributed cognition posits that interaction with the environment implicates and moderates internal representation. For example, Goel (1995) suggests how representation through sketching can challenge a computational theory of mind to indicate a necessity for ambiguity in dealing with ill-defined design problems. Schön (1983) description of the *reflective practice* of design as interaction between the designer and the representation of design intent through drawing and sketching has much in common with distributed cognition. An understanding of a design is constructed insofar as understanding emerges from an interaction with external design representation. By adopting a *reflective practice* as distributed cognition approach, meaning is constructed through reflection, in which design intent emerges from interactions with external design representations.

A number of different terms have emerged within the design literature to describe reflective practice: 'a conversation with design activity' (Lawson 2006, p. 265), 'design talk-back' (Schön 1983, p. 79) as well as 'visual design thinking' (Goldschmidt 1994, p. 160). These terms attempt to capture a definition of design practice that centres on a situated act of designing, whereby representation and reflecting upon solution candidates is critical. In this way, reflective-practice suggests that design ideas towards solution possibilities emerge from the external representation of intent to evaluate the potential of solution candidates. This is an iterative, constructed process in a cognitive sense, whereby reflections stimulate further expressions, triggering iterations of further reflection and decision making. This is shown in Azim Jameel's work in Figures 2.10 and 2.11 where different variations of Study Sketches were developed for the design of a storage system.

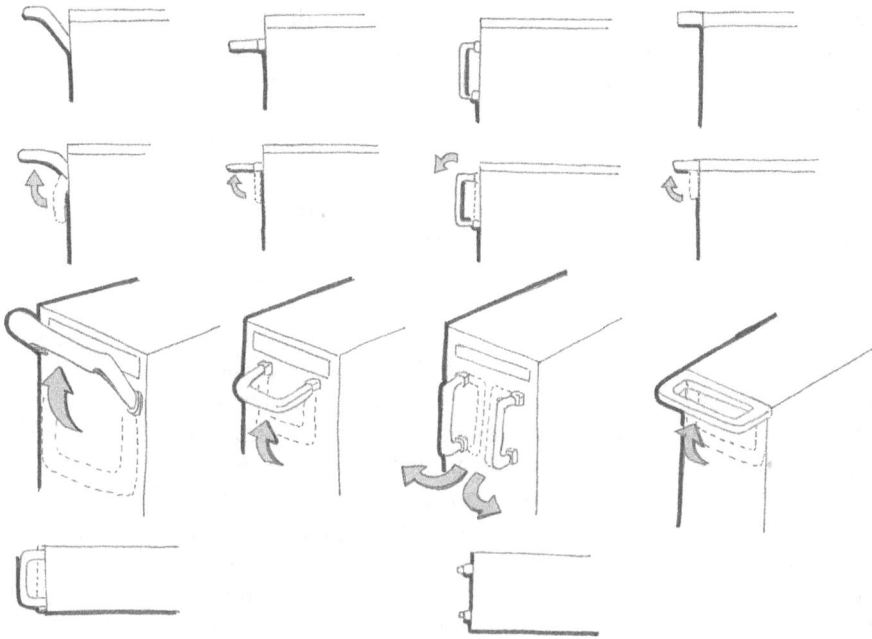

FIGURE 2.10
Study Sketches for a storage system (Jameel, n.d.).

FIGURE 2.11
Study Sketches showing further investigation (Jameel, n.d.).

The roots of *reflective practice*, as a paradigm for understanding design activity, can be traced back to Schön's (1983) study analyzing conversations between a designer and his student. This analysis focused upon design representations used to explore possibilities regarding the position, orientation and design of a building. Using a protocol analysis of conversations between the instructor and the student, Schön (op cit.) describes what he termed conversations between the *materials of the situation*, where the materials are the design representations themselves. Schön (op cit.) goes on to provide a definition of design representation, 'sometimes he makes the final product; more often, he makes a representation – a plan, program or image of an artefact to be constructed by others. He works in particular situations and uses particular materials' (1983, p. 78).

Since design is concerned with the search for more desirable futures, design representation is a method through which intention is expressed, explored and developed. The use of various media enables representations approximating the form and function of a design at greater or lower levels of fidelity. Design representations provide a space for reflection-upon-action to support identification and refinement of solution possibilities (Figure 2.12). Design representations, and the practitioner's reflections upon them, also provide momentum for the evolution of design intent in what Schön terms, 'design moves'. These moves iterate through design representation in a progression towards commitment, in which 'there is a continually evolving system of implications within which the designer reflects-in-action' (Schön 1983, p. 103).

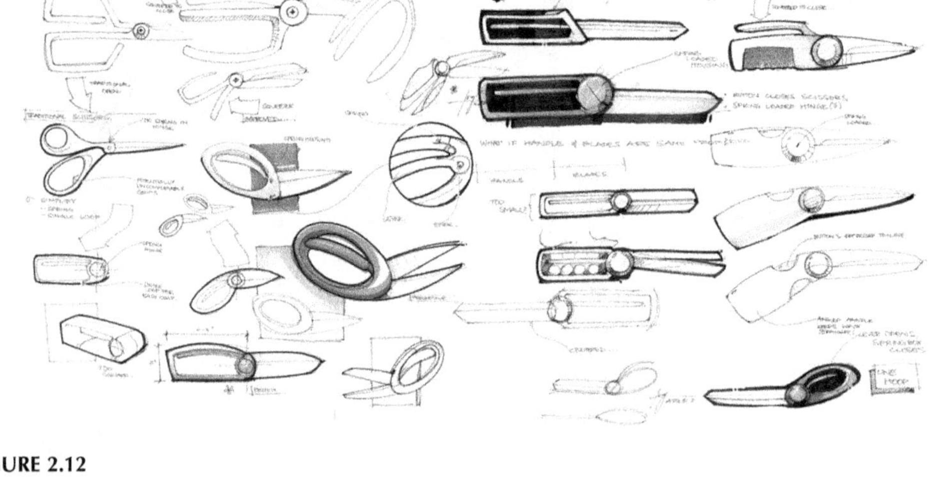

FIGURE 2.12
Idea Sketches and Study Sketches combined (Nugent, n.d.).

The notion of a conversation has been further developed by others; notably from the work of Lawson (2004, p. 84), who describes a two-way conversation through which the designer employs design representation. Lawson adopts a more media-centric approach to consider the ways in which these conversations inform an exploration of the design problem through the tangible embodiment of solution ideas as Sketches as well as CAD tools. His discussion of representation through 3D CAD indicates the influential role that media plays. In particular, Lawson is critical of the ways in which 3D CAD may be used to support a certain type of conversation with a design situation as 'a halting clumsy process that more closely resembles the assembly of a sentence in a foreign language' (Lawson 2004, p. 70). In this sense, Lawson's treatment of *reflective-practice* is oriented towards understanding the influence of the media of design representation. In contrast, Schön's work (op cit.) is focused on the interaction between representation and reflections as a means to understand design.

Like Lawson, Dorta et al. (2008) points to the reflective nature of design practice with a focus on the influence of design embodiment and how embodiment may relate to the changing requirements of a design process. Dorta et al (ibid) discuss a process whereby designers use externalised models of proposed design solutions to interact with mental images. The study indicates how Sketches, Drawings and physical Prototypes are used as a bridge between design thoughts and design activity. This then informs a process where the externalised design embodiment of ideas result in unexpected developments. This then has commonality with Schön's (op cit.) *reflective practice* paradigm for understanding design activity. Both Dorta and Schön indicate a relationship between design representation and design activity, with the designer's idiosyncratic use of representations influencing the kinds of reflections being made. In this sense, an interaction effect between external representation and internal cognition appears necessary to design cognition.

However, although a foundation has been established to understand the use of design representation as part of the reflective practice of design, this approach still does little to tell us how and why design representation is critical to design from a cognitive or psychological point of view. Providing a holistic answer to this question goes beyond the scope of our book. However, work that bridges what we know about design representation with cognitive psychology is fertile ground with the potential to advance both disciplines. For example, examining relevant theoretical constructs within cognitive psychology through the lens of design representation may start to provide a foundation for a theory of design representation. Drawing upon works within cognitive sciences, the following sections explore two concepts and their relation to the kinds of thinking engaged during reflective-practice through design representation.

2.2.2 Design representation and ambiguity

A conventional definition of ambiguity describes a condition of being open to more than one interpretation. Ambiguity is also associated with vagueness, uncertainty and obscurity in the sense that in ambiguity a situation, phenomena, artefact or communication is in some way unclear. By this definition ambiguity has negative connotations; an ambiguous situation may lack detail and clarity. However, in adopting the term as a construct to help our understanding of design representation, the definition of ambiguity, or the condition of being ambiguous, requires a positive grounding. Adopting a *reflective practice* approach to understanding the role and use of design representation, ambiguity can be described, within a design context, as the quality of being inexact, and by doing so providing a greater opportunity for interpretation during reflection.

Design representations can be more or less exact, often dependent upon their purpose, and the stages in the design process they are used. Applied to design representation, ambiguity refers to the extent to which interpretation of meaning may be present during *reflective practice* between internal mental image and external design representation. Ambiguous representation may be further characterised by three types of indeterminacy: *generality*, *vagueness* and *ambiguity*. Generality occurs when an idea that may be descriptively precise specifies a category with many examples. Black (1936), in a criticism of vagueness in language, describes generality as indeterminacy whenever there exists wide plurality in examples related to differences when specifying structure, form or colour. An example of chair is provided in that, while the term may be exact in its description of a product for sitting, generality exists in its application to describe a variety of different chairs used for different purposes. Ambiguity occurs when a choice has not yet been made between two or more alternatives. The Sketches by Azim Jameel for his storage system show ambiguity where the design solution at this stage has not fully materialised (Figure 2.13), whereas the solution is shown to be more defined in Figure 2.14. In Figures 2.15 and 2.16, it can be seen that his designs are more formalised and defined with specifications that include dimension lines and other detail.

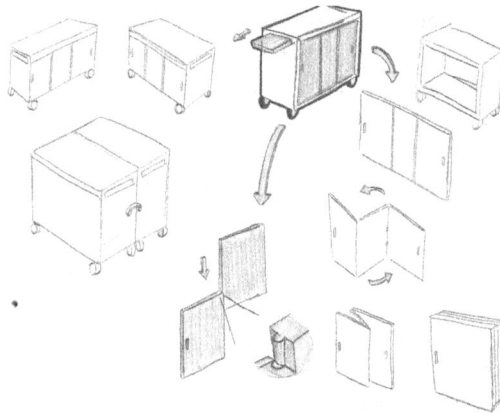

FIGURE 2.13
Idea Sketches for the Smart Box concept (Jameel, n.d.).

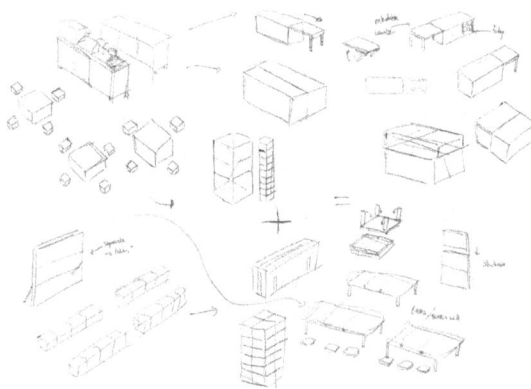

FIGURE 2.14
Another example of Idea Sketches for the Smart Box (Jameel, n.d.).

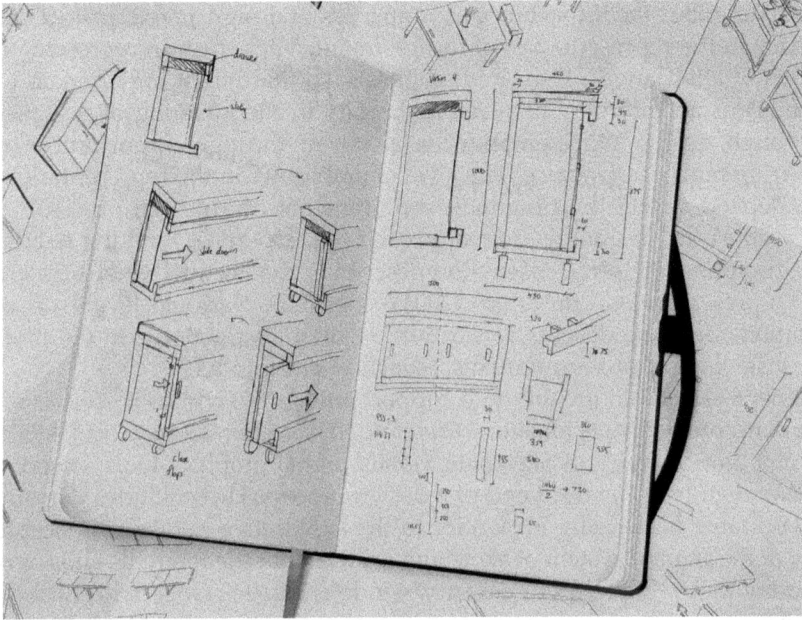

FIGURE 2.15
Information Sketches of the Smart Box (Jameel, n.d.).

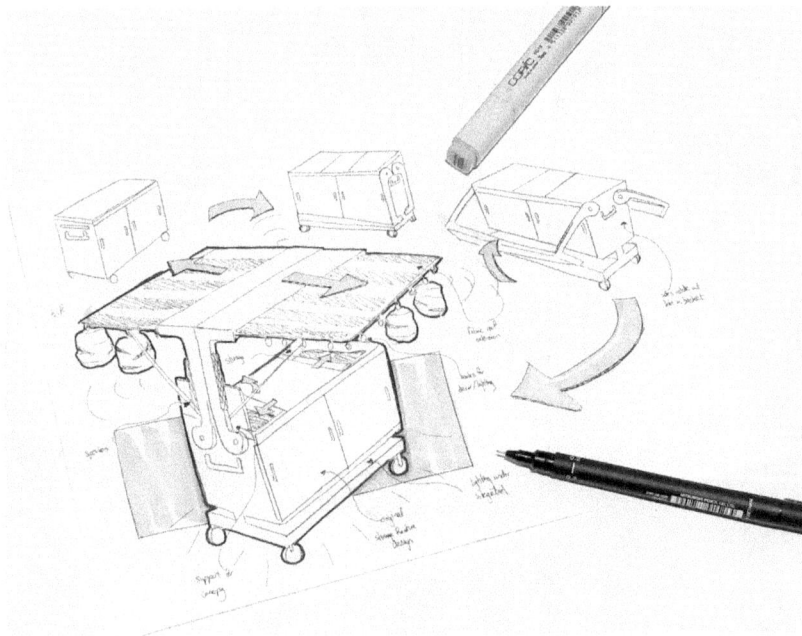

FIGURE 2.16
An Information Sketch with annotations in use (Jameel, n.d.).

Fish (1996) describes the importance of vagueness in design as related to a necessity for indeterminacy in the representation of design intent. Vagueness in representation allows consideration of alternatives, Fish (ibid) suggests, so that intent may remain open to interpretation. Thus an interplay between ambiguity in the meaning of design intent, expressed through design representation, is achieved through opportunity to interpret meaning during reflective practice. The level of ambiguity in the representation of design intent also depends upon the pragmatic requirements of a design process, the type of project, as well as the designer's own approach (i.e. expertise, working habits, working principles, discipline, etc). In a discussion describing the abstract versus concrete nature of design representations, Brereton (2004) considers various levels of ambiguity, 'Representations describe designs at various levels of abstraction. On the more abstract end of the scale lie lists of requirements, Sketches, and Models'.

The designer's sketch has an ability to support ambiguity, and so is seen as effective for use during conceptual design ideation (Brereton 2004; Visser 2006). Goel (1995, p. 193), in an early, comparative study of sketching versus digital graphics tools, describes the importance of ambiguity at the front end of a design process. He concludes that ambiguity in design embodiment is critically important to the explorative nature of conceptualisation, 'Ambiguity of the symbol system of sketching ensures that the referents and/or contents of symbols during the early phases of design are indeterminate'. Goel (ibid) further describes ambiguity as important in avoiding the early crystallization of ideas'. For example, in Figure 2.17, the Coded Sketch by Saeka Hayashi uses symbols to represent meaning in her work, and the coloured regions in the Multi-view Drawing by Wai Lim Chan in Figure 2.18 represent different functional components of the product.

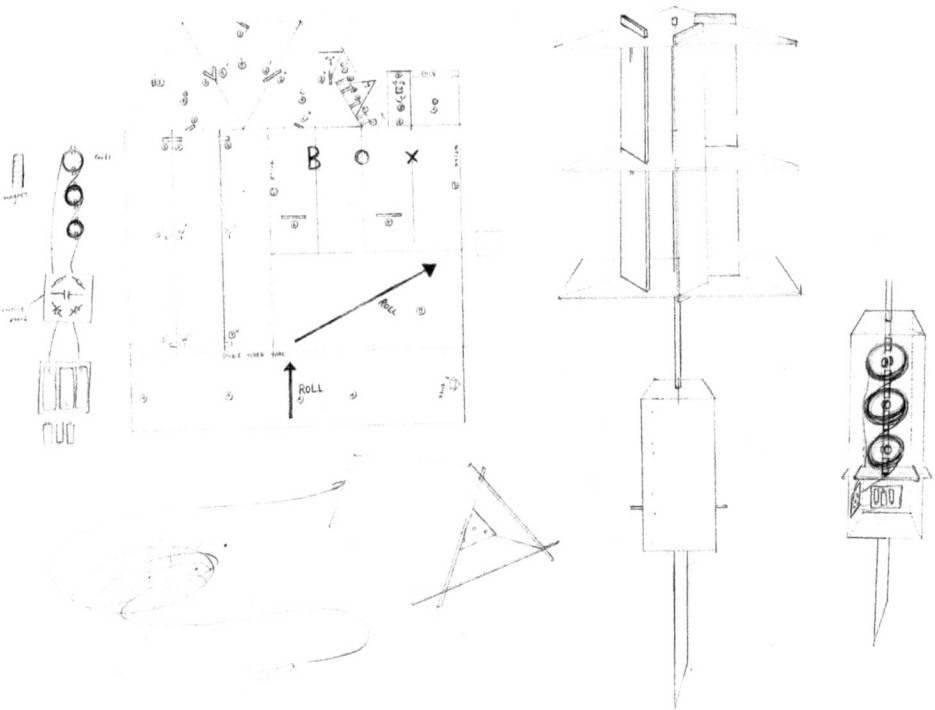

FIGURE 2.17
An example of a Coded Sketch (Hayashi, n.d.).

05 ORION
General Dimension & Section views

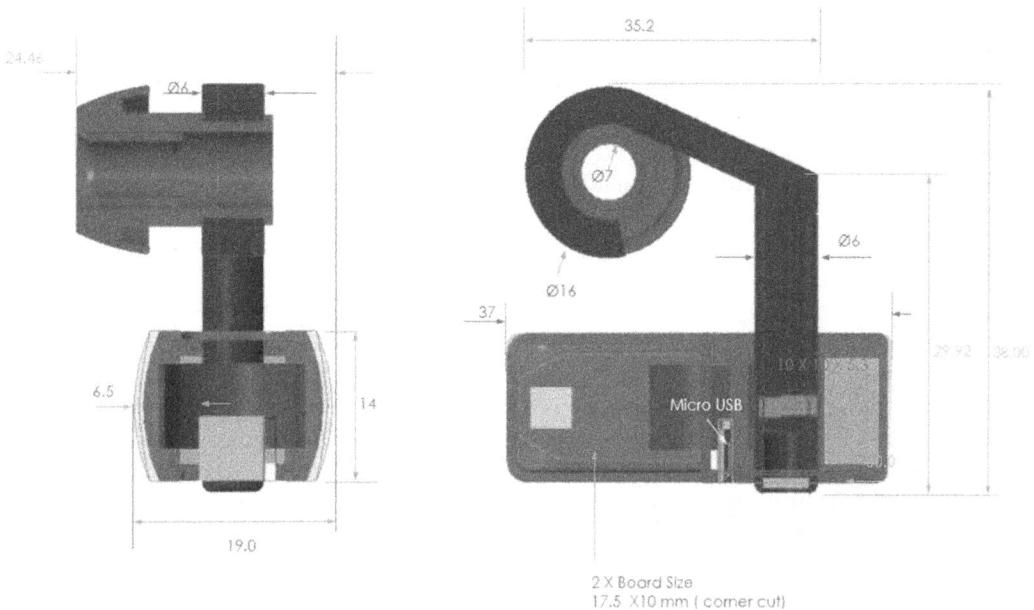

FIGURE 2.18
An example of a Multi-View Drawing (Chan, n.d.).

While we do not claim these dimensions of ambiguity are exclusive, we aim here to provide an operational definition of ambiguity in design representation. Our approach in using ambiguity as means to describe the role and use of design representation as part of a reflective-practice of design rests on the premise of a linear interaction effect between ambiguity and subjective interpretation of design intent. That is, the more ambiguous the representation, the more space for interpretation, where interpretation is often an important characteristic of concept design, providing opportunity for exploration. For example, although Sketches appear rough, inaccurate or incomplete, it is a valuable form of output by providing order and tangibility, while the elements of ambiguity stimulate reinterpretation as an important catalyst in creative transformation of information. The speed of sketch generation also assists in cognitive processing (Garner 1999). Different interpretations due to ambiguity can sometimes be helpful as it could lead to lateral transformations where the design can be reframed to suggest new ideas and alternatives (Garner 2006).

2.2.3 Design representation and fidelity

If ambiguity is associated with an interaction effect between the character of a design representation and the constructed interpretation of meaning, fidelity refers to the degree in which a representation approximates the detail of design intent; how much information is made available and how prescriptive it is. Fidelity has often been used within the literature to describe how close a Prototype, as design representation, is in its form and/or function to a final intended design. For example, how close is the form or surface treatment to the intended final industrial design? How well does it approximate its functional aspects? The further away a representation is, the lower the level of fidelity. Conversely, with increases in fidelity, a design representation moves closer to the form or function of an intended product. Fidelity may also be defined as existing on more than one dimension of interest within a design representation. For example, a Prototype may express form and CMF (colour, material, finish) of a product concept with a high level of fidelity. In other words, it may look very close to the final design outcome. However, the Prototype may not function at all. It may not work like the final intended product solution. Fidelity then, different from ambiguity, learns towards the character of the design representation, rather than interaction effect between representation and interpretation of meaning.

Where ambiguity relates to an imprecise expression of intent, providing opportunities for interpretation, fidelity describes more specifically the distance between a design representation and intentions towards a final design outcome in terms of the detail of the design. And where ambiguity focuses to a greater extent upon the designer's (or stakeholders') interpretation of meaning as representations are reflected upon, fidelity refers to the representation itself, and the gap between a representation's articulation of intent and a final design outcome at project end. The spontaneous Sketches from Spencer Nugent show a dynamic, freestyle approach approach, illustrating the product form from different perspectives and also with varying levels of detail (Figures 2.19 and 2.20).

The representation of information at a lower level of fidelity is useful during concept design to support the generation and exploration of design ideas (Goel 1995; Pipes 2007). Brereton (2004, pp. 83–103) suggests, 'they do not force the designer to pay attention to details that the designer is not yet ready to consider. Different representations make different kinds of information available' (op cit, p. 86). Designers may use low-fidelity representations afforded through the use of sketching to remain open to alternatives or to develop different ideas towards aspects of a design concept before selecting a direction. The Study Sketches and Prototypes created by Oscar Diaz show Sketches and physical manifestations with a varying degree of fidelity to investigate the design idea (Figures 2.21 and 2.22).

FIGURE 2.19
Study Sketches investigating the form of a product (Nugent, n.d.).

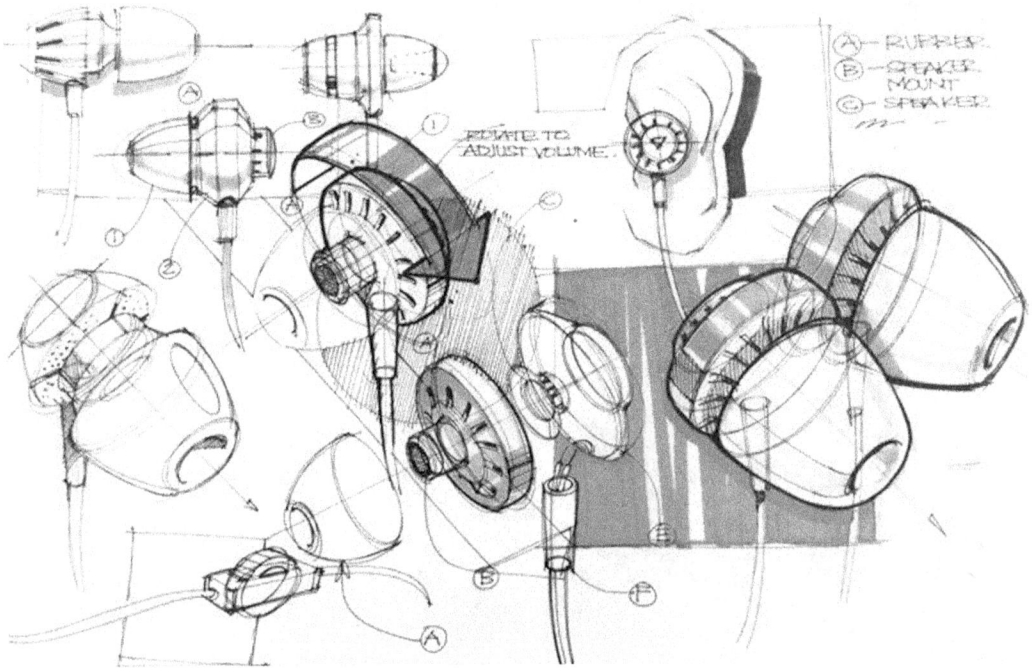

FIGURE 2.20
Information Sketches (Nugent, n.d.).

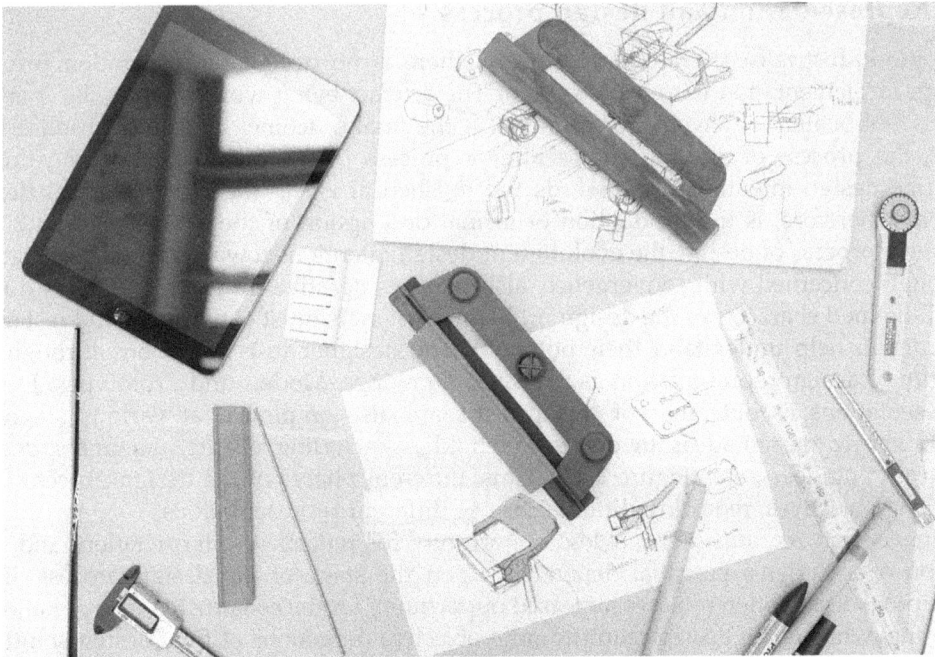

FIGURE 2.21
Study Sketches, Models and Prototypes (Diaz, n.d.).

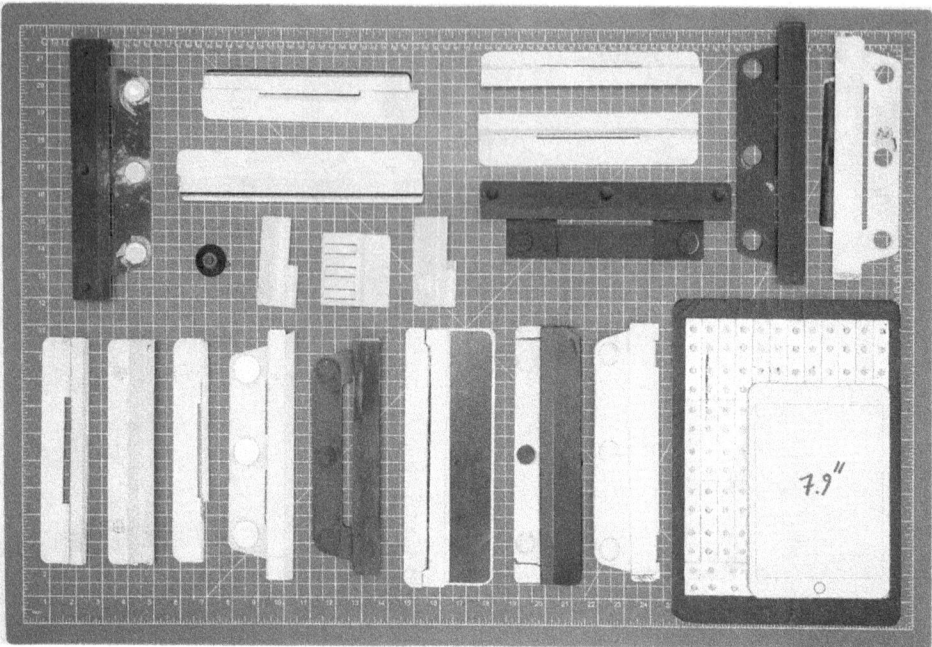

FIGURE 2.22
Iterative improvements of Models produced during the design process (Diaz, n.d.).

2.3 Representation and design process

In a pre-industrialised, craft-based society, there is no design representation prior to design implementation (Cross 2008, p. 3). The artefact being worked upon, be it an artefact, tool or item of jewelry for example, *is* the final outcome. In an industrialised society, the process of making comes after a process of design (ibid). Design aims to describe design intentionality towards the finished artefact, 'the most essential design activity, therefore, is the production of a final description of the artifact' (Cross 2008). During a process of design, the evolution of design intentions towards final specification, although concerned with convergence, also requires iterative divergence. This is due to the ill-defined character of the design problem. Designers must experiment and test ideas in order to help understand their potential. The designer tests intentions through the activity of design representation as Sketches, Drawings, Models and Prototypes. Design representations as outcomes of this process are also employed at various stages to communicate design ideas to other stakeholders including clients, managers, manufacturers, engineers, etc. (Figure 2.23). Thus, different phases of the design process have differing objectives, requiring different use of different representations.

Thus, design requires more or less opportunity for reflection, interpretation, and evaluation of a design's potential dependent upon the stage of the design process. Early conceptual design often requires increased opportunity for the constructed interpretation of meaning, while detail design requires more objective definitions of final design solutions.

FIGURE 2.23
Design meetings and discussions (Prototypum, n.d.).

Mapping the needs of the design process onto the concepts *ambiguity* and *fidelity* in design representation described above, a necessity for increased fidelity and reduced ambiguity follows the design process through *Concept Design*, *Concept Development* and *Detail Design* phases. Adopting this approach, we then define the relationship between design process and the character of design representations as a linear effect. Ambiguity in representation decreases as fidelity increases. This is due to the requirements of phases in the design process and a necessity to articulate design intent unambiguously for production.

2.3.1 Design representation: Concept Design

Concept Design, an initial phase in the design process has been described as an open-ended, fuzzy design ideation stage. This is because design is often initiated by an exploration of ideas to identify an appropriate design solution to take forward to development. In the evaluation of the potential appropriateness of a design idea, the designer attempts to identify a problem-solution pair (Dorst 2015). This exploration and evaluation of alternatives has a relationship to the degree of fidelity in design representation and the level of ambiguity provided in the expression of design intent. This is because, as concept design is characterised by the exploration of possibilities in an attempt to identify more appropriate design solutions, space for subjective interpretation during reflection upon design representation appears important.

That is, ambiguity in the expression of design intent is a necessary component of *Concept Design*. This is because ambiguity provides increased opportunity for the interpretation of design intent. Interpretation may then lead to exploration of alternatives by accommodating differences in opinion, different ideas and reflections on various solution propositions. This is important because conceptual design, characterised as a search for more appropriate solution ideas, necessitates reduced fixation in favour of exploration.

Typical deliverables for a conceptual design phase include approximate descriptions of the form, function and features of the designed artefact (Ulrich and Eppinger 2003); often through design representation. For example, sketching and sketch modelling are design representations often used to support conceptual design ideation (Cross 2008, 2007; Ulrich and Eppinger 2003), where design solutions are considered by all stakeholders in a process of selection. After which, one or more concepts are taken forward to be developed and refined (Ulrich and Eppinger 2003). As with development and detail design, although concept design requires representations to provide ambiguity in expression of intent, consideration for detail, as well as design convergence, is seen in design representation. The stages in the design process may often also iterate, looping back and forth between *Concept Design* and *Concept Development* phases.

Representations such as Sketches and Drawings are the most common ways to convey and communicate information. However, as data on paper is implicit, it relies heavily on one's knowledge, ability and experience to correctly and accurately interpret the intended meaning (Kalay 2004). In the context of industrial design, some domains may use different terms that express the same idea (Sparke 1996). Conversely, the same term may have a different meaning for each discipline. This is particularly the case during conceptual design where design ideas, and their representation, remains necessarily ambiguous and open to interpretation of a proposed solution's (Figures 2.24 and 2.25).

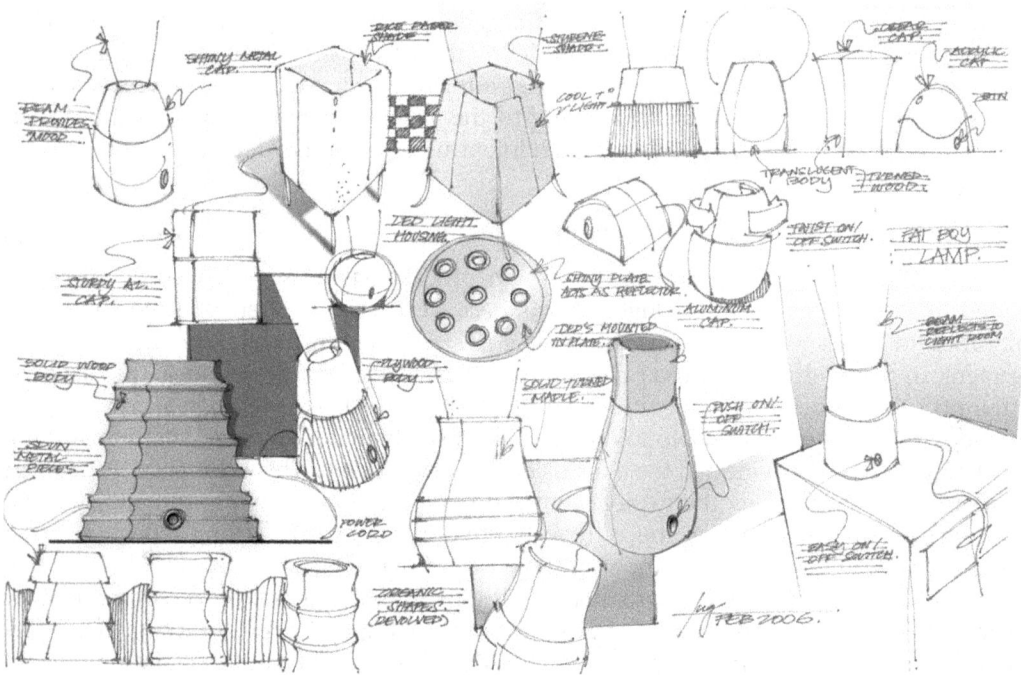

FIGURE 2.24
Various Study Sketches (Nugent, n.d.).

FIGURE 2.25
Another example of Study Sketches (Walters, n.d.).

2.3.2 Design representation: Concept Development

If conceptual design is concerned with the identification of a potential design direction, *Concept Development* moves to bring a chosen direction into greater focus. The process pivots towards furthering an understanding of the detail of a design. Form and function aspects, together with other contextual, production and costing criteria inform the direction of development. Due to the changing nature of the design process, ambiguity starts to recede, while fidelity increases. The vagueness of intent is reduced as design is expressed in a more concrete way. This then reduces opportunity for the interpretation of ideas. Thus, a receding of ambiguity in representation supports the refinement of intent through increased representational fidelity. Although consideration of alternatives is still possible, this is more often related to certain aspects of a proposed design, rather than a greater change in the holistic direction of the design proposition.

Fidelity of representation, on multiple dimensions of interest (i.e. form, function, use, materials, manufacture etc.), begins to increase as intention is expressed at a closer distance to the final design outcome. For example, representation may be characterised by the articulation of a more specific aesthetic form, such as colour, material, finish; function such as technical functionality, mechanical design and part engineering; as well as user considerations including detailed user-interaction elements and interaction design. In doing so, alternatives are reduced as the design process moves to develop design intent in greater detail with increased fidelity on multiple dimensions. Although the design process is increasingly concerned with convergence towards final specification, the process will also alternate between periods of iterative divergence and detailed convergence as design activity progresses towards the final specification of a design solution.

A characteristic of development design is its increasingly structured approach to representation as design becomes increasingly concerned with the definition and communication of more specific design details: 'communication of design information becomes increasingly important as the design becomes more detailed' (Figures 2.26 and 2.27) (Cross, p. 130). Press and Cooper (2003) describe communication of design detail in the increasing use of digital models to communicate design intentions in parallel with the evolution of design specifics.

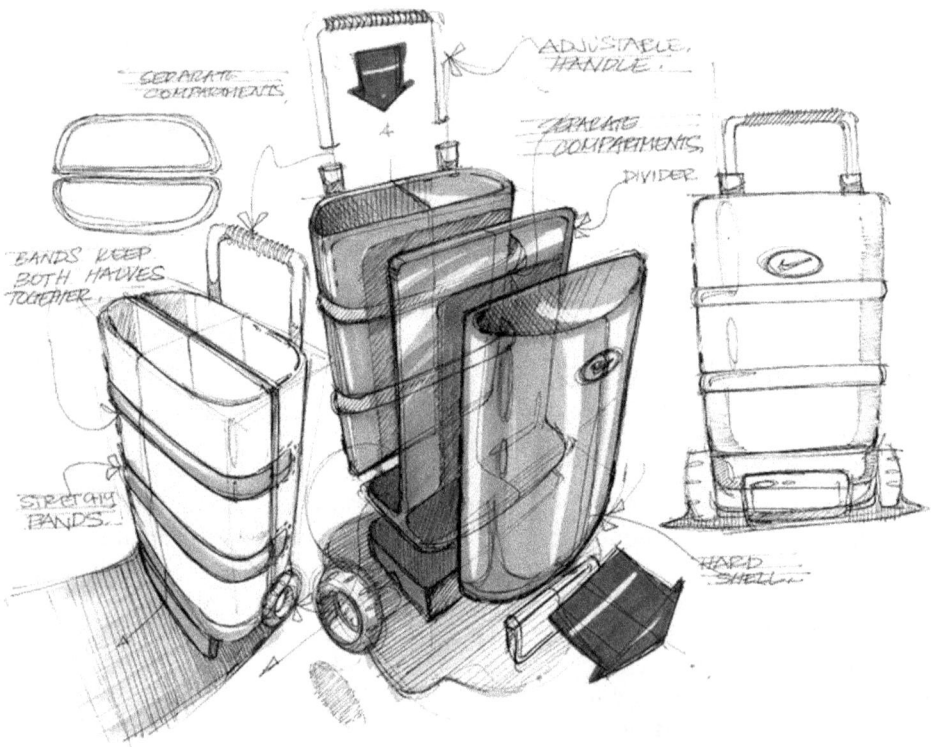

FIGURE 2.26
Information Sketches (Nugent, n.d.).

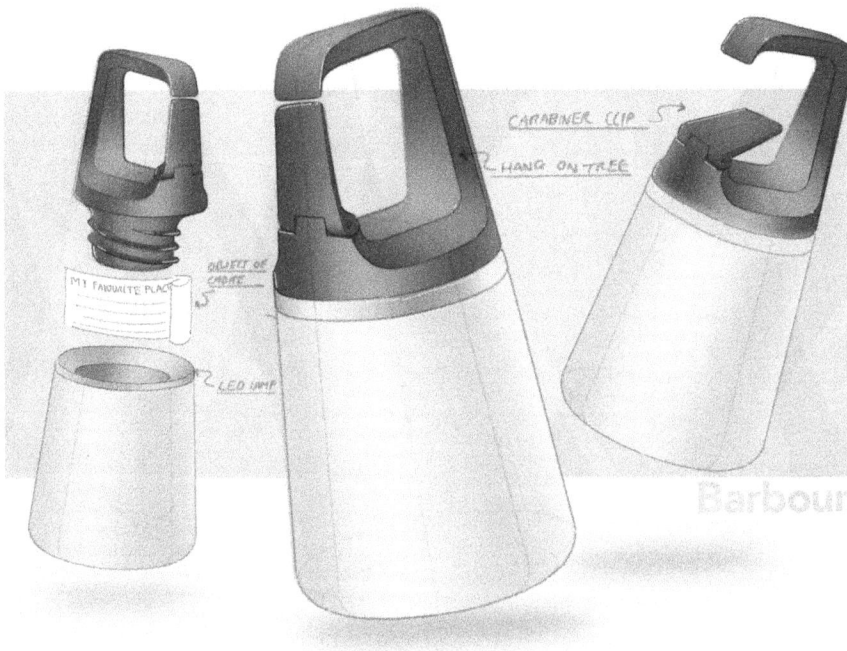

FIGURE 2.27
Renderings of a product (Walters, n.d.).

During *Concept Development*, the importance of design representation starts to slide from representation providing space for subjective construction of meaning through ambiguity, to representation as objective communication of intent derived from increased fidelity. As the purpose of design pivots from a need for interpretation, offered through ambiguity and lower-level fidelity, to a requirement for communication, so the conceptual distance between representation and intended design outcome reduces.

2.3.3 Design representation: Detail Design

During *Detail Design*, the expression of design intent turns to the prescription of product detail. Design representation is characterised by close proximity to the final design solution. At this stage, the opportunity for constructed interpretation is removed. *Detail Design* is concerned with the prescriptive communication of intent in the unambiguous representation of design. Perhaps due to the evolving requirements of this later stage, the kinds of representation used during *Detail Design* may often conform to prescriptive rules of articulation (i.e. Engineering and Parts Drawings, Pre-production Prototypes). These representations aim to eradicate any ambiguity for the purpose of complete definition (Figures 2.28–2.30). Design turns towards the finality of form and technical function prior to manufacture. The prescriptive nature of *Detail Design* can also be seen in Engineering Drawings, that unlike the designer's sketch, work on rigid rules of definition and interpretation.

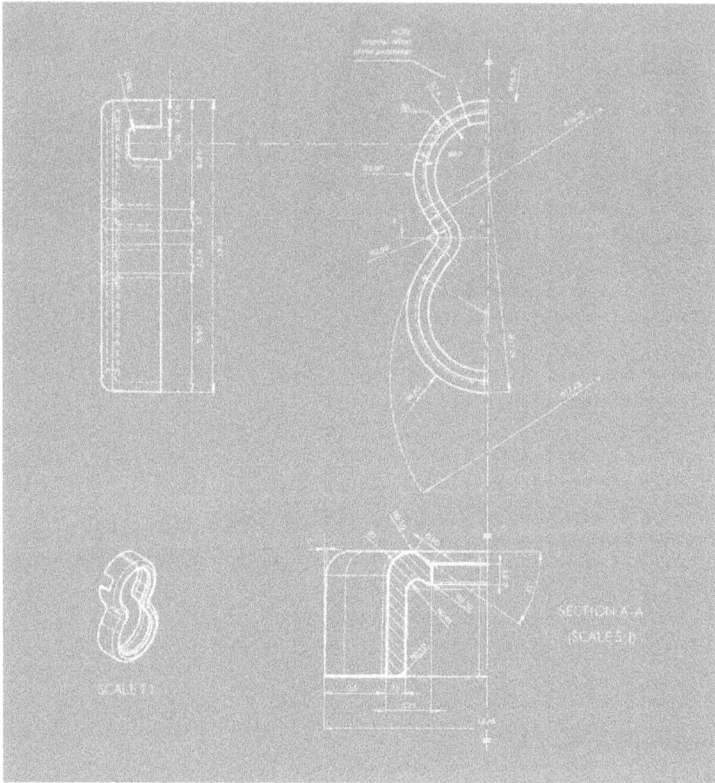

FIGURE 2.28
Multi-View Drawings (Dermachi, n.d.).

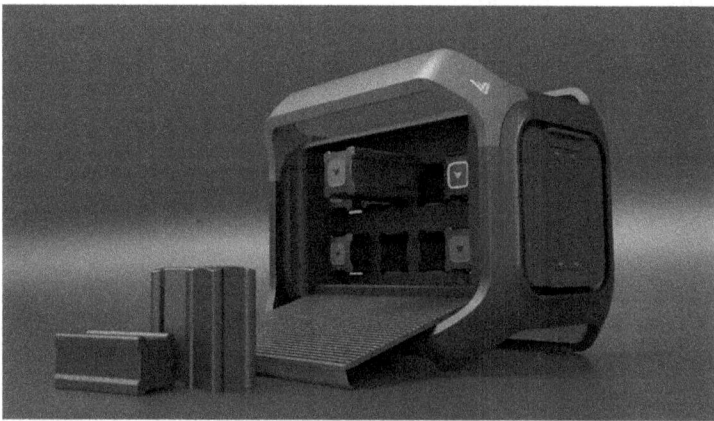

FIGURE 2.29
An example of a product Rendering (Chan, n.d.).

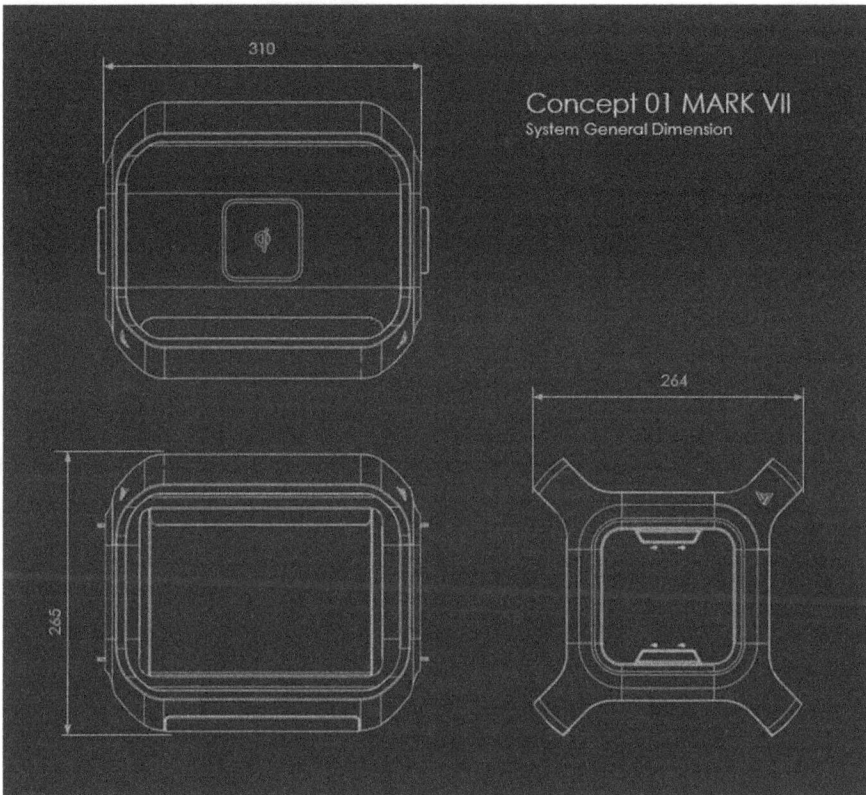

FIGURE 2.30
Multi-View Drawings (Chan, n.d.).

Evidence of a requirement for reduced ambiguity at this phase of the design process is offered by Stacey and Eckert (2003), who caution that ambiguity may have adverse effects in hand-over situations. Their studies show that when design representations that include incomplete information are submitted, recipients interpret according to their own experience and end up with designs that do not reflect the original intent. According to Stacey and Eckert (2003), unintentional ambiguity may arise because of misread codes, contradicting values and missing information; and also occurs when notational conventions are in conflict.

In a separate study to identify the perceived level of technical content or form, Engelbrektsson and Soderman (2004) revealed that hand-made Sketches received the lowest score because of their high level of uncertainty and vagueness. In contrast, virtual reality and rapid prototyping provided higher levels of technical content and form. An increased fidelity of expression resulted in representations being scored as expressing more appropriate designs. Visual design representations, used among developers or between several stakeholders, has also been termed as *intermediary objects*, (Vinck and Jeantet 1995), a *coordinate artefact* (Schmidt and Wagner 2002), or *boundary object* (Star 1989; Maier et al. 2007). They retain their primary purpose across the organisation, yet still allowing use within each discipline. In this sense, representations take the form of artefacts, language and expressions of intent (Wenger 1998; Boujut and Laureillard 2002).

FIGURE 2.31
Design meetings using representations to communicate ideas (Prototypum, n.d.).

Therefore, members need to be clear about the intent and nature of the representation. In addition, as different viewpoints exist among stakeholders, members interpret an object differently or select different aspects from the same representation (Visser 2007; Self 2019). This may be particularly problematic during conceptual design, where representations remain necessarily ambiguous to provide opportunity for interpretation. However, it may just as well become problematic in *Detail Design* if individual stakeholders are unaware of the rules and conventions existing to inform a reading of the design representation, as may be the case with Technical Drawings and detailed illustrations (Figure 2.31). In this sense, an unambiguous representation is dependent upon an agreed understanding of how the representation should be read. Still, in detail design, the intention remains to remove any ambiguity of expressions.

So far, we have positioned ambiguity and fidelity as constructs to explore relationships between design representation and the changing requirements of a staged approach to understand the design process. We have also touched upon a contradiction about the need to accommodate constructed meaning, while at the same time, to avoid misunderstanding that arise from differences in interpretation. The chapter's final sections will discuss design representations that contradict the particular nature of design problems and their solutions.

2.4 Representations, design problems and solutions

Design problems can be described as ill-defined or wicked in their complexity. This is because design as an applied and often multi-disciplinary activity requires the consideration of many variables. Further, a final design solution may potentially take an infinite number of forms dependent upon the way in which the designer both frames the initial design problem and the methods and approaches applied to explore and develop potential solution candidates. This process of problem framing and solution ideation has been described as requiring appositional reasoning (Cross 2011) between problem definition and solution ideation, as a symptom of the design problem's complexity. Rather than a simple progression from problem to solution, design will often involve iterations between, the framing of a design problem (analysis), and the proposition of potential solution candidates (synthesis). In order to achieve these iterations, the designer must look both forward and backward. Backward in a desire to understand the existing. Forward to the proposition and evaluation of potential solution ideas.

Dorst (2011) describes this particular type of design thinking as abductive reasoning, whereby the solution and means to arrive at it cannot be known at the start of a design process. Through the appositional reasoning required to frame an ill-defined design problem (looking back and looking forward), design representation is positioned as a means to facilitate iteration between the design problem and its solution. Through the proposition of solution ideas as design representations, the designer bridges between the design problems and potential solution candidates. Thus, the designer may also moderate both the design process (i.e. concept, development, detail design) and the types of representations used (ambiguity, fidelity) in response to the particular framing of the design problem. Like relations between ambiguity, fidelity and process, the extent of appositional reasoning may also relate to the type of representation used. Our assumption then is that representations high in ambiguity are best placed for reasoning between problem definition and solution ideation (i.e. understanding the problem through solution attempts). As ambiguity recedes and fidelity increases, appositional reasoning reduces in favour of synthesis and refinement as shown in Figure 2.32 where Phase 1 seeks

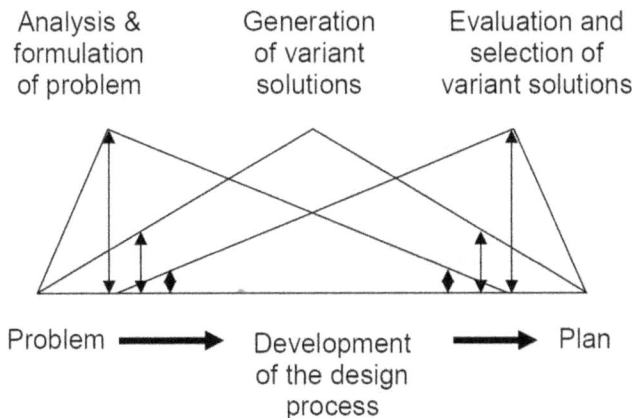

FIGURE 2.32
Boekholt's (1985) model of the design process.

the formulation of the design problem; Phase 2 focuses on the generation of solutions; and finally Phase 3 seeks to evaluate the solutions that have been generated to select a final design solution.

2.5 Design representation: a definition

Despite the gaps in our understanding of design representation, it is important to apply what we know to formulate a definition of the term. In this chapter, we have discussed the character of design representation through the two theoretical constructs, ambiguity and fidelity. As a first attempt to operationalise these two concepts, we then considered how the changing character of representations relates to the process of design as moving from explorative conceptual design, through development and into detail design. The chapter has included a brief account of appositional reasoning between design problems and their solutions. We also provide the foundations for our definition of design representation.

First, design representation is the expression of design intent towards potential solution candidates. That is, design representation's purpose is always to express and communicate ideas towards possible solutions and their related problems (Figure 2.33). Here the aim is to articulate, through representation, the various aspects of a potential design solution (i.e. form, function, use scenarios, technical aspects, manufacturing, etc.). These expressions of design

FIGURE 2.33
Model making in a typical design studio (Shah, n.d.).

intent will be necessarily more or less ambiguous in their communication, at a lower or higher level of detail and fidelity dependent upon stage in the design process.

Second, within the scope of our current discussion, we restrict our definition of design representation to iconic expressions of intent that attempt to approximate the potential design on one or more dimensions of interest. For example, the colour, material and finish of a design, the design's technical function, its form factors or use. Our definition excludes written and spoken language, diagrams as well as visual schematics that do not approximate the attributes of a design in their expression (e.g. theoretical models). In this exclusion, we do not deny the use or importance of these modes of expression and communication, but rather aim to refine the scope of definition to focus upon a particular type of representation in design. Here we restrict our concern for representational expressions as Sketches, Drawings, Models and Prototypes.

Third, design representation is a type of distributed cognition (between design thinking and design representation) that provides unique opportunities to first identify, then explore and finally develop potential solution candidates in response to ill-defined design problems.

2.6 Summary

In this chapter, we have positioned design representation within its historic and epidemiological contexts (Section 2.1). We have examined the role and use of design representation as it relates to Schön's (1983) paradigm of reflective design practice (Section 2.2). Within this discussion, we positioned the two theoretical constructs: *ambiguity* (Section 2.2.2) and *fidelity* (Section 2.2.3) as possible candidates to identify relationships between the character of different design representations, their role and use in design cognition and during a process of design.

Ambiguity was described as related to the designer or stakeholder's response to a design representation, and the degree to which constructed interpretation may be required or designed in the interpretation of meaning. Fidelity, on the other hand, was positioned as a character of the design representation itself. Higher levels of fidelity being closer to a final description of design form and/or function.

Design representation was then considered in terms of its relation to and use within a phase process model of design that moves from *Concept Design*, through *Concept Development* and on to *Detail Design* (Section 2.3). The types of different design representations employed in support of these phases in the design process were then considered and discussed. Although often iterative in nature, we considered the relationship between *ambiguity* and *fidelity* in the expression of intent through representation, and a relation to the changing requirements of process as design moves from conceptual ideation to detailed specification.

The chapter concluded with a short discussion of representation's relation to reasoning between problem definition and solution synthesis in design practice.

Finally, we positioned a working definition of design representation in practice through three broad criteria aimed at defining the unique qualities of design representation. First, design representations aim to express and communicate design intent towards possible solution candidates and the related problems they address. We also limit the scope of our definition to iconic representations that approximate a final design, its material or functional qualities, rather than representation through written word or oral presentation. Third, design representation is a unique type of distributed cognition that provides an opportunity to engage in reasoning (design thinking) between ill-defined design problems and their potential solution. The role and use of design representations will evolve through a process of design. However, these three tenets of design representation: expression of design intent through iconic representation to provide opportunity for reasoning between problem and solution, do not change.

References

Baxter, M. (1995) *Product Design: Practical Methods for the Systematic Development of New Products*. London: Chapman and Hall.

Bilda, Z. and Demirkan, H. (2003) An Insight on Designers' Sketching Activities in Traditional Versus Digital Media. *Design Studies*, Vol. 24(1), pp. 27–50.

Boujut, J.-F. and Laureillard, P. (2002) A Co-Operation Framework for Product-Process Integration in Engineering Design. *Design Studies*, Vol. 23(5), pp. 497–513.

Brereton, M. (2004) Distributed Cognition in Engineering Design: Negotiating between Abstract and Material Representations. In: Goldschmidt, G. and Porter, W. (eds), *Design Representation*. London: Springer, pp. 83–103.

Cross, N. (1992) *Design Principles and Practice Block 1: An Introduction to Design*. Milton Keynes: Open University Press.

Cross, N. (2007) *Designerly Ways of Knowing*. Basel: Birkhauser.

Cross, N. (2008) *Engineering Design Methods: Strategies for Product Design* (4th ed.). Chichester: John Wiley and Sons.

Cross, N. (2011) *Design Thinking: Understanding How Designers Think and Work*. London: Bloomsbury.

Dorst, K. (2011) The Core of 'Design Thinking' and Its Application. *Design Studies*, 32(6), pp. 521–532.

Dorst, K. (2015) Frame Creation and Design in the Expanded Field. *She Ji: The Journal of Design, Economics, and Innovation*, Vol. 1(1), pp. 22–33.

Dorta, T. Perez, E. and Lesage, A. (2008) The Ideation Gap: Hybrid Tools, Design Flow and Practice. *Design Studies*, 29(2), pp. 121–141.

Eckert, C. and Boujut, J. F. (2003) The Role of Objects in Design Co-Operation: Communication through Physical or Virtual Objects. *Computer Supported Cooperative Work* (12), pp. 145–151.

Evans, M., Wallace, D., Cheshire, D. and Sener, B. (2004) An Evaluation of Haptic Feedback Modelling During Industrial Design Practice. *Design Studies*, 26(5), pp. 487–508.

Fish, J. C. (1996) How Sketches Work: A Cognitive Theory for Improved System Design. PhD Thesis. Loughborough University of Technology. Quoted in: Do, E. Y., Gross, M. D., Neiman, B. and Zimring, C. (2000) Intentions in and Relations among Design Drawings. *Design Studies*, 21(5), pp. 483–503.

Garner, S. (1999) Drawing and Designing: An Analysis of Sketching and Its Outputs as Displayed by Individuals and Pairs When Engaged in Design Tasks. Ph.D. Thesis. Loughborough: Loughborough University.

Garner, S. (2006) *Modelling Workbook 1: T211 Design and Designing Workbook 1 Technology Level 2* (2nd ed.). Milton Keynes: The Open University Press.

Goel, V. (1995) *Sketches of Thought*. London: The MIT Press.

Goldschmidt, G. (1992) Serial Sketching: Visual Problem Solving in Designing. *Cybernetics and Systems*, Vol. 23, pp. 191–219.

Goldschmidt, G. (1994) On Visual Design Thinking: The Vis Kids of Architecture. *Design Studies*, Vol. 15(2), pp. 158–174.

Goldschmidt, G. (1997) Capturing Indeterminism: Representation in the Design Problem Space. *Design Studies*, Vol. 18(4), pp. 441–455.

Jonson, B. (2002) Sketching Now. *The International Journal of Art and Design Education*, Vol. 21(3), pp. 246–253.

Jonson, B. (2005) Design Ideation: The Conceptual Sketch in the Digital Age. *Design Studies*, 26(6), pp. 613–624.

Kalay, Y. (2004) *Architecture's New Media*. Cambridge: MIT Press.

Lawson, B. (2004) *What Designers Know*. Oxford: Architectural Press.

Lawson, B. (2006) *How Designers Think: The Design. Process Demystified*(4th ed.). Oxford: Oxford University Press.

Maier, A. M., Hepperle, C., et al. (2007) Associations between Factors Influencing Engineering Design Communication. *International Conference on Engineering Design ICED '07, 28–31 August 2007*, Cite Des Sciences Et De L'Industrie, Paris, France.

Nelson, G. and Stolterman, E. (2003) *The Design Way: Intentional Change in an Unpredictable World*. New Jersey: Educational Technology Publications.

Pipes, A. (2007) *Drawing for Designers*. London: Laurance King Publishing.

Press, M. and Cooper, R. (2003) *The Design Experience*. London: Ashgate.

Rittel, H. W. J. and Webber, M. M. (1973) Dilemmas in a General Theory of Planning. *Policy Sciences*, Vol. 4, pp. 155–169.

Robertson, B. F. and Radcliff, D. F. (2009) Impact of CAD Tools on Creative Problem Solving in Engineering Design. *Computer-Aided Design*, Vol. 41(2), pp. 136–146.

Schmidt, K. and Wagner, I. (2002) Coordinative Artifacts in Architectural Practice. In: Blay-Fornarino, M., Pinna-Dery, A. M., Schmidt, K. and Zaraté, P. (eds.), *Cooperative Systems Design: A Challenge of the Mobility Age. Proceedings of the* Fifth *International Conference on the Design of Cooperative Systems, COOP 2002*, pp. 257–274.

Schön, D. (1983) *The Reflective Practitioner*. London: Ashgate.

Self, J. (2019) Communication through Design Sketches: Implications for Stakeholder Interpretation during Concept Design. *Design Studies*, Vol. 63, pp. 1–36.

Söderman, M. (2002) Comparing Desktop Virtual Reality with Handmade Sketches and Real Products: Exploring Key Aspects for End-Users' Understanding of Proposed Products. *Journal of Design Research*, 2(1), pp. 7–26.

Sparke, P. (1996) *An Introduction to Design and Culture in the 20th Century*. London: Allen and Unwin Publishers. Quoted in: Hague, R., Campbell, I., Dickens, P. and Reeves P. (2002) Integration of Rapid Manufacturing in Design. *Time Compression Technologies*, Vol. 10(4), p. 48.

Stacey, M. and Eckert, C. (2003) Against Ambiguity. *Computer Supported Cooperative Work*, Vol. 12, pp. 153–183.

Star, S. L. (1989) The Structure of Ill-Structured Solutions: Boundary Objects and Heterogenous Distributed Problem Solving. In: Gasser and Huhns (eds), *Distributed Artificial Intelligence*(Vol. 2). London: Pitman, pp. 37–54.

Stolterman, E., McAtee, J., Royer, D. and Thandapani, S. (2008) Designerly Tools. *Proceedings of DRS2008, Design Research Society Conference*, Sheffield University, UK, 16–19 July.

Tovey, M. and Owen, J. (2000) Sketching and Direct CAD Modelling in Automotive Design. *Design Studies*, Vol. 21(6), pp. 569–588.

Tovey, M., Porter, S. and Newman, R. (2003) Sketching, Concept Development and Automotive Design. *Design Studies*, Vol. 24(2), pp. 135–153.

Ulrich, K. and Eppinger, E. (2003) *Product Design and Development* (Int. ed.). New York: McGraw-Hill Education.

Vinck, D. and Jeantet, A. (1995) Mediating and Commissioning Objects in the Sociotechnical Process of Product Design: A Conceptual Approach. In: MacLean, D., Saviotti, P. and Vinck, D. (eds), *Management and New Technology: Design, Networks and Strategy.* Brussels: COST Social Science Series.

Visser, W. (2007) Collaborative Designers' Different Representations. *International Conference on Engineering Design (ICED '07),* Cite Des Sciences Et De L'Industrie, Paris, France, 28–31 August.

Visser, W. (2006) *The Cognitive Artifacts of Designing.* New York: Routledge.

Wenger, E. (1998) *Communities of Practice - Learning, Meaning, and Identity.* New York: Cambridge University Press.

3

Design representation in practice

3.1 The purpose of design representation

When an idea has been crystallised in the mind, it is usually reproduced as a sketch or physical model through a process of hand-eye coordination. At this point, the image becomes new information (Purcell and Gero 1998; Tovey et al. 2003). The designer perceives and identifies patterns to construct new knowledge and to develop the next idea (Schön and Wiggins 1992). This process, occurring between the designer and the material representation of intent, has been termed 'interactive imagery' by Goldschmidt (1991a). It stops when specifications are met, when creativity has been exhausted or when time and cost become a limiting factor. Visual design representations also relieve the cognitive load from memory to enable further mental processing (Koutamanis 1993; Suwa and Tversky 1997; Schweikardt and Gross 2000; Romer et al. 2001). They serve as an aid to enable ideas to be recalled or to be checked (Goldschmidt 1994, 1989, 1991b; Andreasen 1994; ; Fish 1996; Ulrich and Eppinger 2003; Hendry 2004). Representation supports designers in the assessment of whether the current idea satisfies the project goals as a whole or at a component level (Purcell and Gero 1998; Seitamaa-Hakkarainen and Hakkarainen 2000). As discussed in chapter 2, this interaction between thinking and representation allows testing and modification before investing in manufacture and production (Garner 2004).

Importantly, then, representations are open to extension, modification, and interpretation (Schmidt and Wagner 2004). For example, Van Eck (2015) suggests one of the most important contributions of design representation is to facilitate and validate the next design method to be used in the process. Due to their role as means to validate and modify, representations support the designer's thinking to allow the discovery of new ideas and to stimulate dialogue for questions, insights, revisions and answers (Suwa and Tversky 1996; Tohidi et al. 2006).

Evidence of design representation's role in design thinking is provided by Gorner (1994), who, in an early study of experienced designers, found for 69 percent of respondents sketching was important in the development of design solutions. Similarly, Engelbrektsson and Soderman (2004) performed studies to investigate the types of representations used for design work in general and for communication with customers in particular. Sketches, Prototypes and Construction Drawings were most common, with emergent media (i.e. virtual reality) rarely used. Results also indicated, in communication with customers, Prototypes were most often used, with 3D CAD more popular than Construction Drawings; that were difficult for a non-technical audience. Likewise, Romer et al. (2001) found that sketches were most used for developing solutions and for communication, while CAD was often employed for documentation. Simple models were primarily applied in support of communication; with complex models popular for testing solutions. Related to our own discussion of ambiguity and fidelity of representation (Chapter 2), Garner (2006) observed that designers

DOI: 10.1201/9781003227694-3

generate sketches with different levels of complexity, whereby sketches with less complexity have a higher level of abstraction.

The objective measurement of representational types has also attracted attention as an opportunity to understand their use at different phases of design development. For example, complexity in Sketches may be measured in terms of the type of information shown within the sketch. To this end, McGown et al. (1998) developed a 'complexity scale' that measures a sketch's degree of transformation using qualitative judgements. Monochrome line Drawings with no shading were rated as examples of complexity level one. Complexity level two involves sketches that include rough shading to indicate shapes and forms. Higher complexity sketches were more realistic, included enhanced design details and the use of colour.

Song and Agogino (2004) conducted experiments involving 57 senior mechanical engineering students working on a project-based product design course in the Department of Mechanical Engineering at the University of California. The authors analysed the project journal, drawings and group documents and characterized their findings by means of metrics. The matrix consisted of: (1) Generation (how the drawing was generated and whether it was a new drawing for a new idea or a transformation of a previous drawing); (2) the type of sketch (such as a Thinking Sketch, Prescriptive Sketch, or Talking Sketch); (3) the medium used (such as traditional freehand sketches or in digital form including CAD Drawings, photos, etc.); (4) the type of representation (2D, 3D versus 2D Multi-view Drawings, etc.); (5) the annotation (labels, lists, narratives, dimensions, and calculations); (5) the level of detail (based on the complexity of the sketch).

The Song and Agogino (ibid) study found that the largest percentage of sketches were traditional freehand sketches with a medium level of complexity. CAD Drawings emerged in the conceptual design stage and thereafter greatly increased in the final stages of the design process. They also observed that as the design progressed, students sketched more in 3D and less in 2D. The use of labels and narratives was identified as the primary method of annotation in sketches for all stages. However, such annotations decreased in the later stages of the process, and dimensions and calculations increased. There was a marked increase in the sketches level of detail, which increased increased as the design moved from the preliminary stage to the detailed design. The study confirmed that the total number of sketches, and the use of 3D representations and prescriptive sketches in later design stages, contributed towards a positive effect on the design outcome. Similarly, Figures 3.1 to 3.4 show a series of highly detailed representations of Functional Concept Models and Multi-View Drawings during the design of a bed-side table lamp. The use of 3D Printed Prototypes, illustrations and 3D CAD enabled a closer study of the product details.

From a social perspective, physical representations can also enable the developer to convey information to other stakeholders that might otherwise be difficult to express in words (Eckert and Boujut 2003). In a further study, McKoy *et al.* (2001) provided evidence that graphical representation, such as sketching, is the preferred medium of expression; better-suited for producing good design as compared to words in terms of spontaneity, creativity and quality. Visual design representation, such as sketching, enables others to quickly understand, participate, describe, and contribute towards a project. Through integrating the perspectives of different members, shared representations can create a common frame of reference among stakeholders, allowing continuous comparison of options and to rationalise the design in terms of form and function (Ferguson 1992; Johansson *et al.* 2001; Do 2002; Buxton 2007). This insight is supported by Logan and Radcliffe (2000) who shows that representations, when used individually and collectively, enable common understanding through visual points of reference. Goldschmidt (2007) indicates how sketches allow the

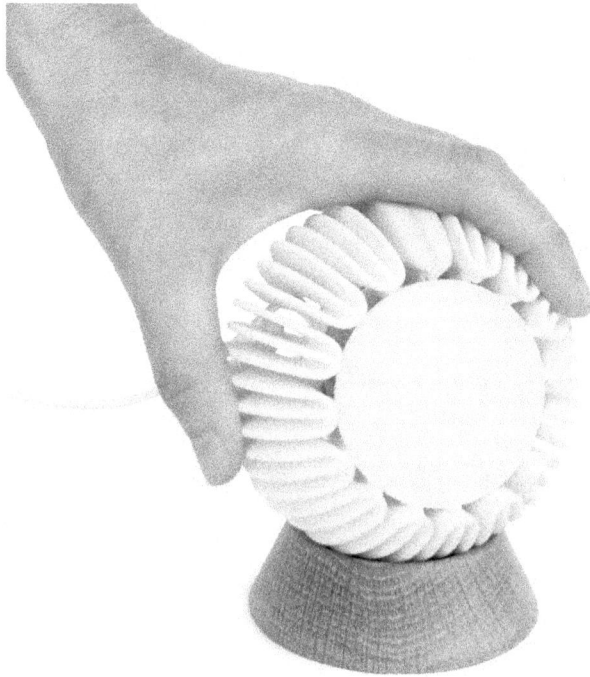

FIGURE 3.1
A Functional Concept Model (Dermachi, n.d.).

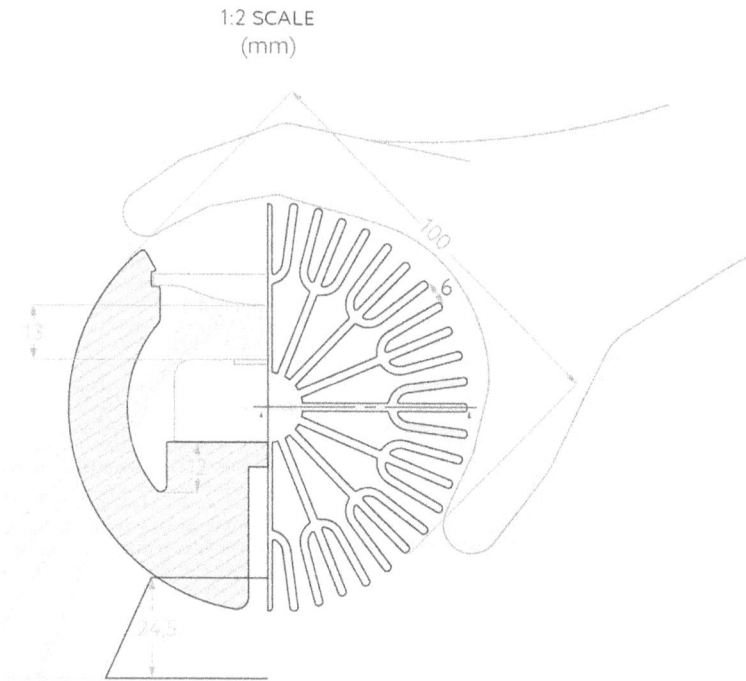

FIGURE 3.2
A Single-View Drawing showing a cross-section view (Dermachi, n.d.).

FIGURE 3.3
A Rendering showing a cut-out view of the product (Dermachi, n.d.).

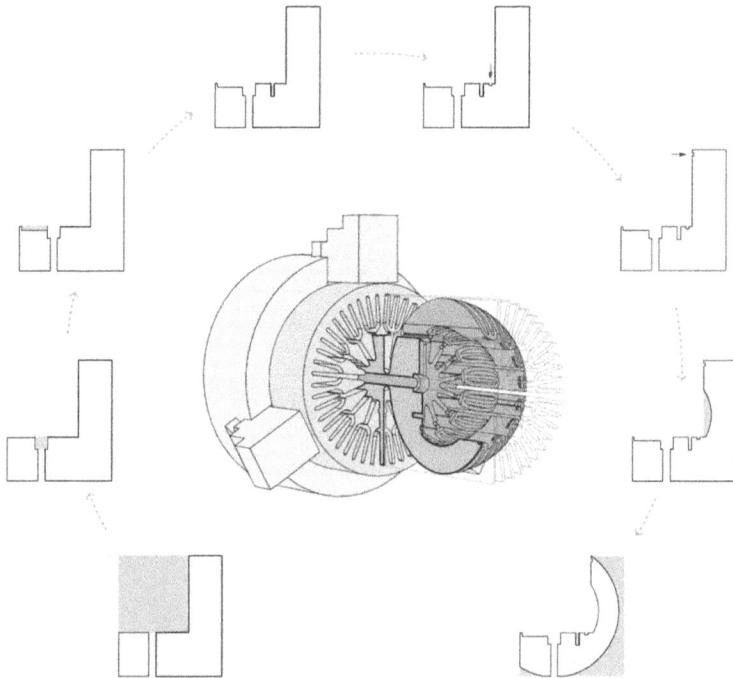

FIGURE 3.4
A Multi-View Drawing of the product (Dermachi, n.d.).

mental models of individual designers to converge for enhanced shared understanding between stakeholders. By acting as a medium for pointing, talking and sketching, sketches
function as mediators; and through manipulation, provide feedback to the person and to the
observer (Heath and Luff 1991; Perry and Sanderson 1998; Gutwin and Greenberg 2002).

As a key element of design activity, representations promote communication, leading
to an enhanced discussion of ideas (Lawson 1994; Scrivener and Clark 1994; Bilda *et al.*
2006). They encourage creative group activities to enable multi-disciplinary members to
share the same attitude towards a project (Leonard-Barton 1991; Schrage 1993; Ulrich and
Eppinger 2003; Olofsson and Sjölén 2005; Alisantoso et al. 2006). They help bridge barriers
between different perspectives and build a platform for sharing ideas, to persuade and
point out issues. Visual design representations also support more effective communication with external stakeholders such as model-makers, contractors and the client. To sum
up, evidence suggests sketches as the focus of interaction, supporting collaborative work
(Lakin 1990; Robertson 1996; Perry and Sanderson 1998; Eckert and Boujut 2003).

While sketches are frequently used by industrial designers in practice, the use of
Computer-Aided Design (CAD) offers different advantages. For example, CAD allows in-
stantaneous storage, quick transmission and the ability to rework designs on demand.
Despite these benefits, manual sketching on paper still presents a much faster and freer
approach without the need to type commands or to specify determined shapes or sizes (Do
2002). In more recent years, the use of tablets or colour-accurate touchscreen displays and
pressure-sensitive digital pens allow designers to sketch directly onto a screen. As a result,
computer images can sometimes be better suited for working in detail as they can be rotated,
moved and visualised realistically (Utterback et al. 2006). In terms of digital 3D modelling,
CAD surface modelling is more commonly used to produce aesthetic detail, while solid
modelling provides technical precision and is used for rapid prototyping (Johansson et al.
2001; Cross 2007) (Figure 3.2). The 3D CAD information can be saved in a proprietary format
to allow further editing or exported in other data formats such as 3D PDF where parts can be
annotated in a 3D workspace. Data formats such as STL, VRML, STEP, 3MF, and AMF are
other formats used by designers and engineers to store 3D CAD data (Figures 3.5–3.9).

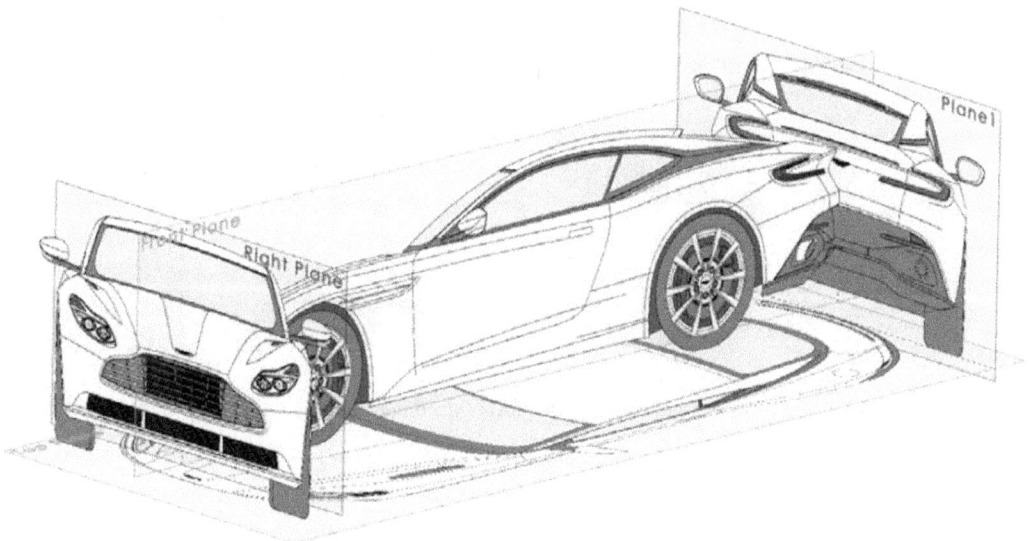

FIGURE 3.5
Using 2D images as reference when creating a 3D CAD model (Mortimer, n.d.).

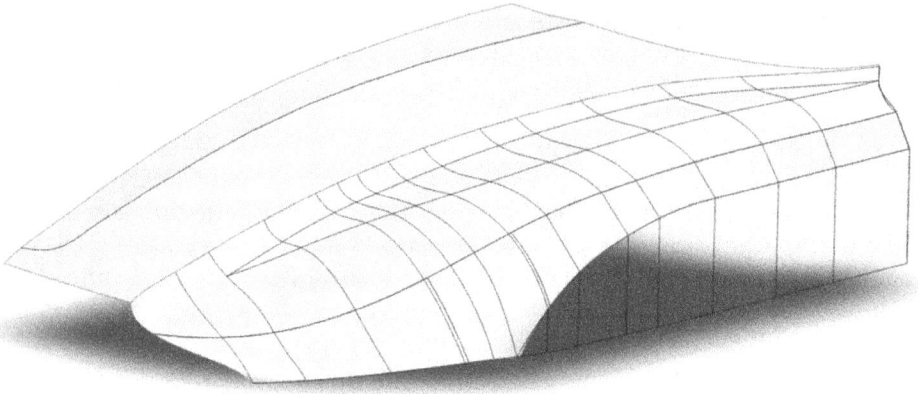

FIGURE 3.6
A surface model generated in 3D CAD (Mortimer, n.d.).

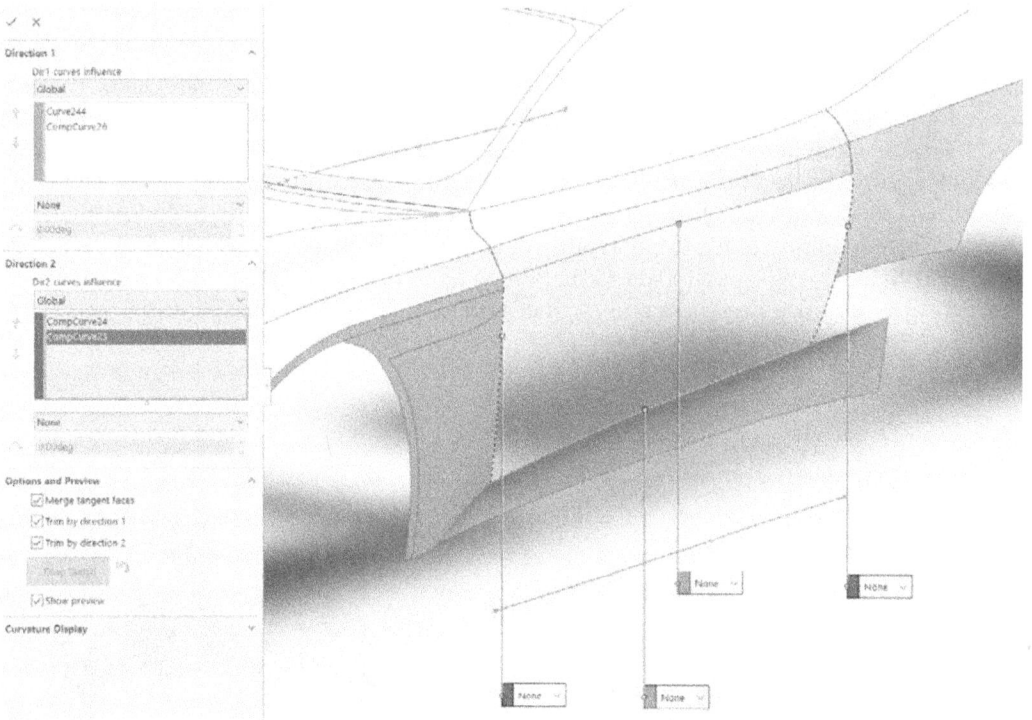

FIGURE 3.7
Another example of a surface model generated in 3D CAD (Mortimer, n.d.).

FIGURE 3.8
An Appearance Model realised using 3D CAD (Mortimer, n.d.).

FIGURE 3.9
An Appearance Model produced using Additive Manufacturing (3D Printing) (Mortimer, n.d.).

Regarding the representation of technical design aspects, Ullman et al. (1990) acknowledged the importance of representations in engineering. These include Technical Drawings or Construction Plans that provide instructions for the production line (Lawson 1997). Other representations such as scaled Drawings allow greater control when managing the magnitude of parts (Jones 1992). Exploded views show overlapping components positioned in a uniform direction and describe component relationships in terms of assembly and manufacture, while structured diagrams illustrate connections, analysis and graphical data (Ulusoy 1999). Towards the later stages of the design process, representations are used to check and detect last-minute errors (Boote 2006).

Returning to the purpose of design representation as influenced by stage in process, Olofsson and Sjölén (2005) noted that sketches and illustrations may be classified according to four main uses: for *investigation, exploration, explanation* and *persuasion*. For example, sketches are used to emphasise form, size, proportion and colour. Sketches may be drawn in different perspectives to explain shapes and connections that would otherwise be limited if seen from only one view. Other benefits in terms of visual understanding include step-by-step illustrations that explain actions, such as how an object would work; or using cross-section lines to describe the shape and form. Of course, various sketches and illustrations may overlap two or more of these categories. However, the Olofsson and Sjölén (ibid) taxonomy is useful in its ability to make more explicit a relationship between the purpose of a design representation, its application at a phase-in process, and how purpose and process interact to influence the type of representation used. For example, at the end of *Concept Design*, representations may be used as a persuasive tool to sell the design concept to management, the marketing team or the client (Tovey 1989; Löwgren 2004).

In another approach to classification of representation, Visser (2007) describes the purpose of representations as grouped around the *personal, social, aesthetic* and *technical*. These categories broadly relate to, on the one hand, representations used to support design thinking (*personal*), or to communicate design intent (*social*). On the other, aesthetic and technical aspects relate to the requirements of the design itself. In a personal setting, representations assist in achieving a clearer mental process, for cognitive distribution including recording, organising, reasoning and discovery. In terms of social aspects, they may aid communication and support group activities, integrate the perspectives of multi-disciplinary members and forge a common frame of reference. Aesthetic aspects are concerned with how a design can be communicated or visualised, while technical aspects relate to the technical or functional details behind the design.

3.2 Types of design representations

In Chapter 2, we discussed the act of design where mental images are visualised and then processed. They are externalised through words, gestures, references and representations that allow concepts and solutions to be formulated (Bly 1988a; Tang 1989; Eckert and Stacey 2000). Representations are objects or things that mean something else (Kaplan and Kaplan 1982; Palmer 1987). According to Saddler (2001), design representations are, 'a perceptible expression of a design idea, proposal or fact'. In line with Johnson (1998), visual design representations are defined as artefacts that reproduce properties of a product by means of a physical or digital form. The most common forms of visual design representation are marks on paper with colour, shading and text (Arnheim 1969; Brown 2003). During sketching practice, various types of mark-making take place on paper. Cooper (2018) observed slashing marks being used to surround and make volume, or the use of repeated curling marks and dots to establish the relationship between objects; or lighter and darker marks used to differentiate levels of space. *Overdrawing* or retracing used to emphasize features within a particular sketch (Cook and Agah 2009; Van Sommers 1984). Designers traced Drawings repetitively, used different tones and added symbols that were part of a process of focusing, selecting, shape interpretation and shape refinement. McGown et al. (1998) further suggested that the use of text captions provide an enhanced understanding of the sketched design. The added use of human figures or *mannikins* provide a better sense of scale.

Ullman et al. (1990) considered three features of Engineering Drawings. Firstly, the type of marks-on-paper. For example, marks supporting notation and graphic representations including textual notes, lists, dimensions, and calculations; and graphic representations including drawings of objects and their functions, plots and charts. Secondly, the medium to produce the drawing, such as traditional versus digital media. Thirdly, the type of graphic representation such as a 2D Drawing versus a 3D Drawing; including the level of abstraction of the information to be represented such as from an abstract concept to the final, detailed design.

Traditionally, paper-based representations have been employed prior to computer augmented representation, thus the term 'pencils before pixels' (Baskinger 2008). The reason given for this approach is to allow designers to think spatially about their designs; to feel and have a physical connection with what they are imagining (Figure 3.10). This connection comes from the hands and bodies acting with the eyes to touch, and make movement (Cooper 2018). This previously identified necessity also relates to the embodied nature of design representation. The expression of and reflection upon design intent through representation, provides opportunity to explore the potential of ideas. In this, representations act as scaffold for design thinking, wherein cognition is distributed between external design representation, and internal representation to provide opportunity to reflect-in-action (Schön 1983). Other modes of representation include producing Scenario Storyboards, Working Prototypes, 3D CAD Models and Virtual Reality (Van

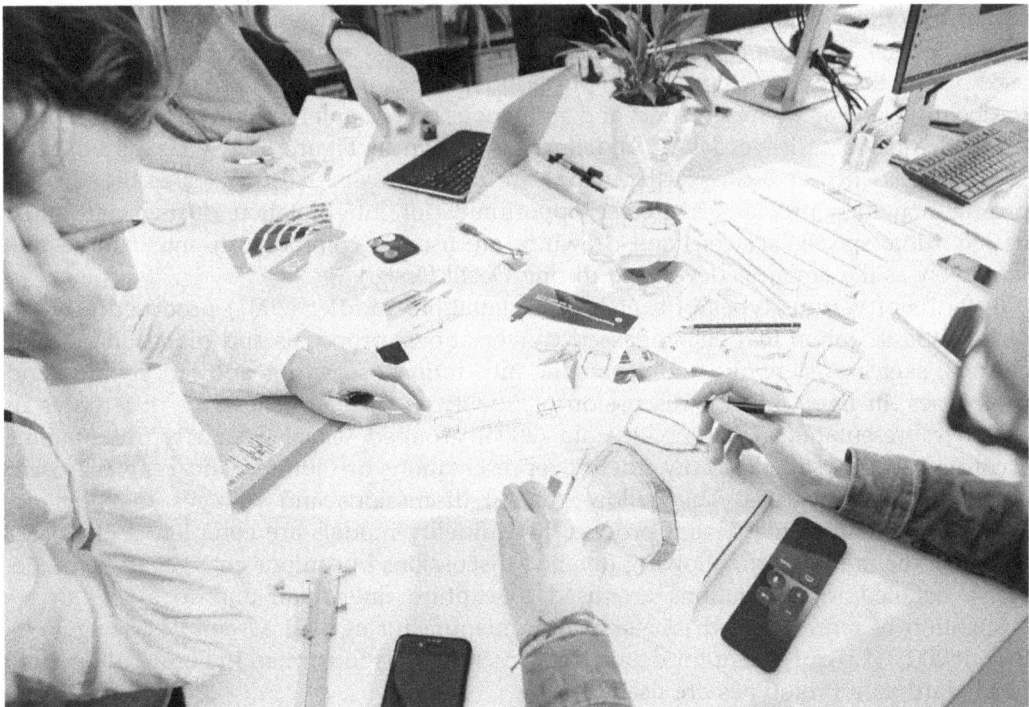

FIGURE 3.10
Design ideation meeting (Prototypum, n.d.).

Welie and Van der Veer 2000; Suri 2003). These can be observed where there is greater realism when moving from 2D to 3D representations (Leonard-Barton 1991). Despite the fact that 3D objects are tangible and more realistic, 2D visual design representations offer minimal commitment, are intuitive and are often easier to create (Lipson and Shpitalni 2000; Holmquist 2005; Cardella et al. 2006). In terms of choice, studies by Johansson et al. (2001), found that large and small companies used a wide range of visual design representations but advanced technologies such as virtual reality are still relatively unpopular as they are often costly and less intuitive.

Bar-Eli (2013) proposed three distinct characteristics of sketching behaviour including a *realization sketching profile*, where the designer uses sketching and writing to develop optimal and detailed solutions. A *learning sketching profile* to develop a variety of options for solutions from diverse points of view. And a *designer sketching profile* that uses abstract sketches to theorize ideas and to tell a personal story. Other researchers have proposed classifications to group visual design representations. Chiu (2002) proposed that Sketches, Orthographic Drawings, tables and photographs could be grouped as being asynchronous, while visual presentations with oral explanations were synchronous. Goldschmidt and Porter (2004) identify four classifications of representations: *internal/ external, transient/ durable, self-generated/ ready-made* and *abstract /concrete*. Goldschmidt and Porter's (2004) taxonomy focused upon the cognitive aspect of representation, and bare most commonality with a discussion of representation as a scaffold for design thinking during a process of design (Chapter 2).

Related to the ways in which design representation informs design thinking, internal representations may be described as residing in the mind while external forms include written lists and drawings. Transient representations such as dialogues and gestures are seldom recorded, while physical representations such as models that are tangible can be stored. Self-generated representations such as conversations and dialogues are transient and used during the design activity, while physical representations consist of materials such as cardboard, wood and wire. Abstract and ambiguous (see Chapter 2) representations such as loose Idea Sketches leave details undefined as shown in Figure 3.11; whereas other representations such as Technical Drawings are more specific. Abstract or ambiguous representation then provides a greater opportunity for interpretation during conceptual design. More specific sketches and drawings are used to communicate intent more prescriptively as the design is developed during Detail Design.

In terms of various types of sketch representation, Saddler (2001) proposed a broad form of classification that encompassed conversations, proposals and plans, spaces and clusters, sketches, symbolic and schematic illustrations, scenarios and storyboards and prototypes. In line with our discussion of fidelity as means to describe differences in design representation (Chapter 2), Cain (2005) grouped visual design representations according to their fidelity. Low fidelity representations are limited and only represent certain product features. They allow general discussions and changes to be made. Towards the end of the design process, high fidelity models are complete and closely resemble the final design, allowing detailed discussions but minor changes to be made. These detailed representations are used to capture fine detail and to allow an investigation of more focused issues such as manufacturing and assembly (Wong 1992; Brandt 2005) A requirement for detailed representation is illustrated in Figure 3.12 where Final Hardware Prototypes are used.

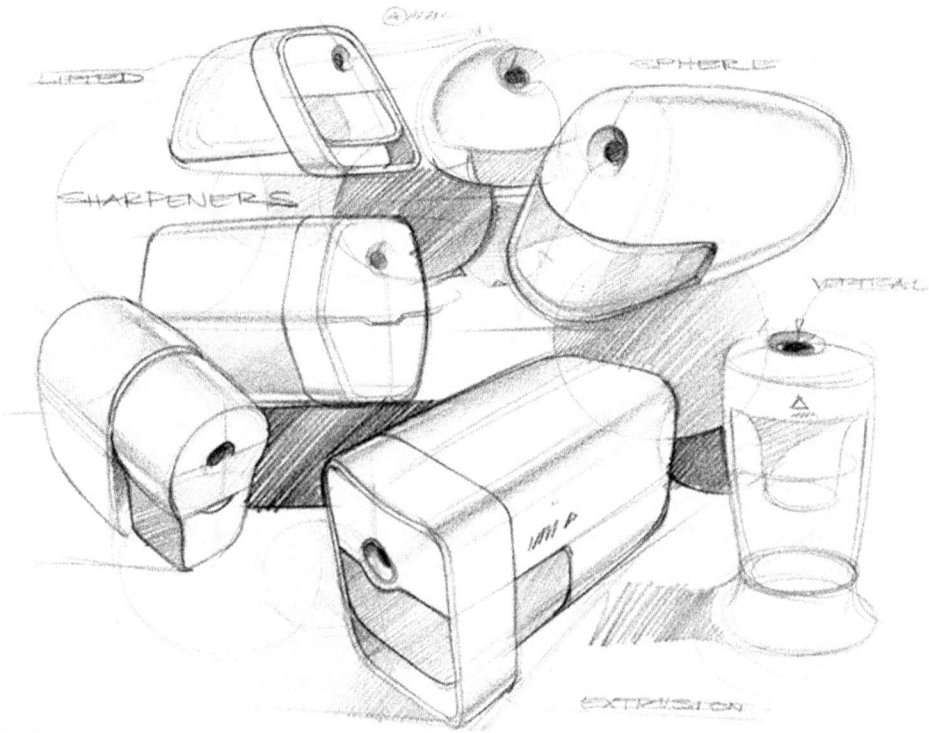

FIGURE 3.11
Idea Sketches (Nugent, n.d.).

FIGURE 3.12
Final Hardware Prototypes (Prototypum, n.d.).

The kinds of tools and processes used in the embodiment of intentions will depend on the purpose and stage of the design process. Sketches and drawings are often used during conceptual design and development stages (Cross 2007, p. 108). Later, to aid specification, models and prototypes are employed (ibid). Potter (2002, p. 16) describes design embodiments as instructions for manufacture that may take many forms, such as detailed Working Drawings, Presentation Drawings for clients and scaled models. Common in the use of all tools is the designer's wish to experiment, test and evaluate design intentions (Cross, op cit, Cross 2008, p. 10). However, in reality, design practice is a complex mix of influences; from technical and manufacturing constraints to delivering the client's needs and the designer's own idiosyncratic methods of working, the company culture, their experience and the working environment (Cross 2007, p. 46).

3.3 Tools of design representations

The use of tools within design practice, in particular tools of design sketching, (Bilda and Demirkan 2003; Fish 2004 ; Jonson 2005), but also 3D CAD (Computer Aided Design) tools (Goel 1995; Dorish 2001; Dorta et al. 2008; Lawson 2004), have received significant attention in design research. However, there is little empirical work to describe the relationship between the designer's attitudes towards the use of tools and how this may impact design activity. A notable exception is Stolterman et al. (2008, p. 4) framework for the study of how practising designers actually view and evaluate tools. Stolterman et al (ibid) explored design tools in terms of the purpose of design practice, the activity required to achieve the purpose and the tool(s) considered by the designers to best support that activity. Findings, less surprisingly, indicate that the designer's idiosyncratic approach to practice does influence their choice of tool and its use (ibid). Less is known, however, about the reasons why differences exist in terms of design expertise. Baber (2003, p. 13) refers to the wider context of tool use as allowing, 'the user to act upon their environment to achieve specific goals'. Use is described as part of a co-dependent activity including the designer, the tool, and the context of design activity which is a requirement of the design process. Even with the emergence of an increasing variety of digital design tools such as 3D CAD, additive manufacturing, graphics tablets, haptic technologies and virtual reality (Evans et al. 2004), sketching is still valued as a core competency by many designers (Pipes 2007, p. 12). Powell (Op cit) described the ability to sketch to support the designer's thought process as having 'a conversation with themselves', and to 'keep pace with the speed of one's own thinking' (ibid, p. 6).

According to Tjalve et al. (1979) and Buur and Andreasen (1989a), there are six key aspects forming a morphology that should be considered before creating a representation. First, properties such as structure, form, material, dimension and surface must be determined. Second, the receiver must be identified in order to decide which sets the criteria to apply. Third, appropriate codes are chosen, such as graphical to convey information for communication including electrical symbols and drafting conventions (section lines, projections, dimension lines, etc.). Next, the technique or the method used to create the representation reflects the quality of the representation. Fifth, tools such as pencils or pens must be chosen. Lastly, the right representation medium such as card, paper or a digital format needs to be selected (Figure 3.13 and 3.14). The use of sketching helps designers record and compare different ideas on paper, while the use of CAD helps them to focus on the more detailed and realistic elements (Shih et al. 2017). Such challenges and benefits of sketching and CAD modelling during the conceptual design phase are

FIGURE 3.13
3D Sketch Models being produced (Shah, n.d.).

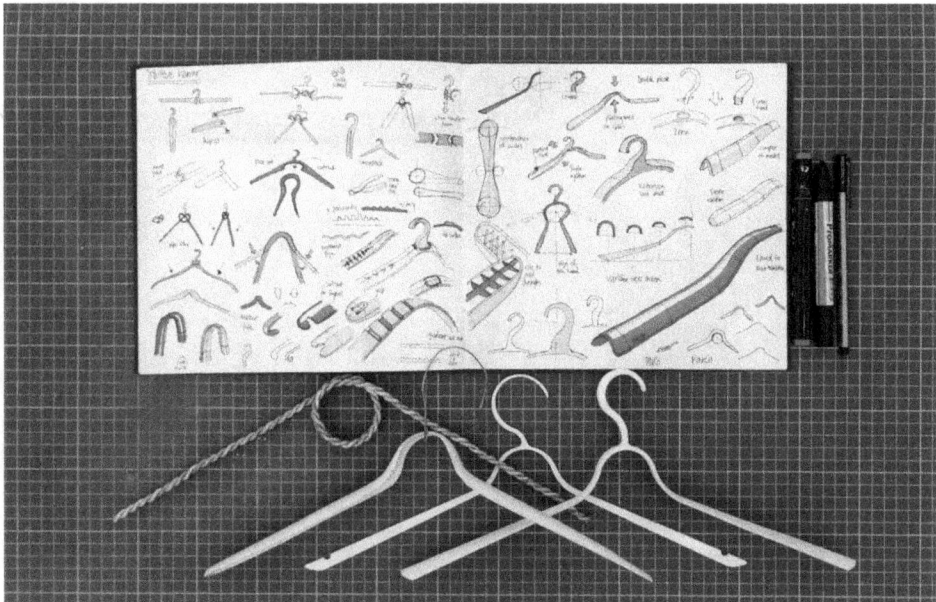

FIGURE 3.14
Study Sketches and Functional Concept Models (Shah, n.d.).

outlined by Rahimian et al. (2008). The benefits of manual sketching include flexibility in the ideation process due to the tangible interface, it is relatively easy to use, easy to learn, easy to change to design alternatives, an ability to use different drawing scales and a possible trade-off between accuracy and clearness, allowing design ideas to be reliably maintained during the design process, as well as to review and compare all documents.

In terms of the drawbacks of sketching, there is less capability to shift from micro to macro level and vice versa, fewer visualisation details, less straightforward for editing or reviewing, more difficult to add or control details of design alternatives, and difficult to transition to other design stages because of the physical nature. CAD, however, offers easier documentation, having the capability for zooming and panning for an easier walkthrough, being able to hide an object or group of objects on the screen, being able to undo and make changes, and to achieve a more detailed, realistic and elaborated perspective due to a much higher capability of visualisation. In terms of challenges when using CAD, gaining proficiency in the programme takes time, having an arduous use of input and output devices that may interrupt the flow of creativity, and also losing the consistency of ideas due to the lack of an ability to control the design idea in an artistic way.

Despite the advantages and limitations of using manual or digital media, there is evidence that some designers prefer to switch between different types of media by alternating the use of sketches and CAD. This is in line with the work of Do (2005), who coined the concept of the, 'right tool-right time'. This refers to design environments that provide the tools that the designer needs, rather than being constrained with a specific tool or media. Bilda and Demirkan (2003) also found that the use of different representational media resulted in different types of design thinking and making, whereby the goals and intentions of designers changed more frequently when using traditional media. Figure 3.15 shows a typical studio where designers work with traditional media as well as with digital tools and software.

FIGURE 3.15
Working on renderings using markers (Choudhury, n.d.).

Self (2012) has also suggested a necessity for the designer to understand the strengths and limitations of sketching and CAD to be best placed in applying them to support design activity at various stages of the design process. In further work Self et al. (2014) identify the concept of *tool-focused* design activity as describing a condition whereby the practitioner is required to explicitly attend to the CAD tool. This may be due to a designer's lack of proficiency in tool use, or a limitation in tool functionality in relation to a particular design or representational requirements.

The term 'medium' in singular, or 'media' in plural, refers to tools and materials where something can be expressed, communicated or achieved (AskOxford 2008). Pavel (2005) commented that the choice of the medium should enable designers to quickly express themselves without losing the design intent. In earlier work, Gantz (2005) commented that today's digital tools have also evolved to support faster development work, providing more accessibility and being more economical. On the other hand, hybrid media has been used to define representations from a mixed origin or composition. They allow the developer to explore complex geometries quickly in a traditional medium and then on the computer to harness the advantages of saving, editing and transferring digital information (Dorta and Pérez 2006). An example of a 2D hybrid media is shown in the concept sketches for the Jordan B'2rue basketball shoes, which use coloured pencils and markers on paper, then scanned and edited in Adobe Illustrator and finished with Adobe Photoshop (Pipes 2007). An example of a 3D hybrid media is shown by Dorta (2005) who conducted iterations between the physical and digital model. After creating a physical model, it was scanned to obtain the digital data that was then manipulated and sent for sdditive manufacturing. Other hybrid approaches proposed by Dorta (2008) include immersive sketching techniques (2D) and immersive model making techniques (3D). As there are no fixed rules regarding the use of media, numerous combinations are possible for a hybrid approach. A square drawn on a piece of paper may be scanned and then digitally transformed into a cube. When this happens, 2D media has been changed into 3D digital media. The developer could either print the image on paper (a 2D media) or have it additively manufactured into a 3D artefact. Because hybrid techniques allow a limitless number of possible combinations, the output of a representation can be classified as manual 2D media, digital 2D media; manual 3D media or digital 3D media.

3.4 Manual 2D media

The term *manual* refers to the act of making or working on something with one's hands as opposed to using digital methods (AskOxford 2008). It is an 'analogue' approach where the qualities or properties of an object are changed by physical methods (Longman Dictionary of Contemporary English 2005). Evans (2002) labelled this as a 'conventional' approach as it does not involve the use of computers. Manual media is used to describe materials and hand equipment used to produce a visual design representation through hand-eye coordination and articulation without computers. A wide range of manual 2D media may be used, including paper, pencils, erasers, pens and markers. In the book *Paper Prototyping*, Synder (2002) promoted the use of paper and other materials due to their speed, low cost and ease of application when creating mockups. By understanding and carefully selecting appropriate media, developers can achieve better results. For example, the colour, type and grain of paper affect the appearance of coloured pencil, ballpoint pen or marker quality (Olofsson and Sjölén 2005). Pencils are available in graphite, carbon or coloured pigments and tones may be achieved by choosing the level of hardness. The B to 6B series is favoured by industrial designers for general sketching, while harder H to 9H types are more suitable for technical and engineering drafting. Other pencils include water-soluble coloured variants that may be brushed with water to produce a washed effect (Pipes 2007) (Figure 3.16).

The use of ball-point, felt-tip and technical pens allow thick and thin lines of ink on paper. Sketching techniques include hatching by shading slanted lines over an area, stippling, or controlling the pen pressure (Olofsson and Sjölén 2005). Markers are used in various colour palettes and are applied in layers to darken an area with other media such as colour pencils and pastel to highlight and enhance the artwork. Other manual tools include charcoal, airbrush, conte crayons, gouache, watercolour and a geometry set consisting of compasses, dividers, rulers, protractors and set squares for precise lines and curves. Other manual tools include compasses, protractors, stencil templates, French curves and bendy splines (Pipes 2007). Coyne et al. (2002) observed that particular equipment, tools, and devices are associated with the designer. For example, a drafting pen is associated with an architect and a chisel is associated with a stonemason as symbols of their trade.

FIGURE 3.16
Idea Sketches (Prototypum, n.d.).

3.5 Digital 2D media

The term *digital* refers to the use of a system in which information is created, recorded or sent electronically by computers (Longman Dictionary of Contemporary English 2005). Digital media is used to describe electronic forms of media created, viewed and manipulated by computers to produce visual design representations. They are based on binary machine language and controlled with a user interface (Kalay 2004). Digital input devices allow the developer to enter and manipulate data on the computer. They include keyboards, digital pens, 2D image scanners, digital tablets and other haptic devices. The digital tablet or a touchscreen computer mimics the use of manual techniques by allowing designers to create digital 2D media through a program on the system. Popular digital tablets include those manufactured by Wacom, as well as the iPad by Apple Inc. Accessories include digital art markers with a chisel-tipped nib to perform a rendered stroke, or digital airbrushes that simulate a pressure-based ink application.

In terms of software, vector graphic editors enable geometrical lines, curves and shapes to be drawn, based on mathematical algorithms. When the image is scaled, no pixelation occurs and image quality is not compromised. In contrast, vector graphics are more suitable for line art, illustrations, technical diagrams and flowcharts as they are not able to produce realistic images. A common vector graphic editor is the use of Adobe Illustrator. Raster graphic editors enable the creation, editing, manipulating and viewing of photo-realistic images. Adobe Photoshop is a popular raster graphic editor. Other forms of raster editors are natural-media drawing programs that allow natural sketching on digital tablets with a pressure-sensitive pen that includes Autodesk SketchBook. Certain types of software allow information to be arranged into layers, where each layer may contain certain attributes. Kurtoglu and Stahovich (2002) proposed a program that takes freehand sketches as an input. The symbol recognizer (geometric reasoner) remembers the individual symbols and physical reasoning identifies the meaning of the symbol. Key to this tool is that it is able to make meaningful connections between the symbols in different contexts such as in mechanical and electronic diagrams. Their system suggests that freehand sketches are inaccurate and ambiguous; and clarity can be achieved by reasoning the overall context rather than examining each part of the sketch in isolation. Digital methods also allow traceability where reference points provide electronic records of the development of the design. Users are able to refer back to previous versions of the drawing (Coyne et al. 2002). It is also interesting to note that the use of new devices introduces new terminologies, and new languages to describe practices (ibid). For example, the use of the drafting pen introduces words such as line width, precision, sharpness, texturing, etc. Another example is the use of CAD that introduces terms such as manipulating, adding and subtracting volumes. Such devices can also borrow words used in other areas such as 'zooming' which is an optical term.

Research has found that CAD-based tools are highly effective to support tasks such as analysis or manufacture (Woodbury and Burrow 2006). The use of digital industrial design tools and techniques have the potential to reduce the time from product planning to product sale. Early research carried out by Tovey (1997) in computer-aided car styling showed that digital techniques could support the cognitive processes of creative transformation (such as ambiguity) but also gave rise to a very different nature of output. Digital tools encourage efficiency gains, leading to faster product development and enhanced design output (Chen and Owen 1998). Computer-based sketching methods provide several advantages when compared to paper-based methods, such as for storage, meaningful processing, are more durable, and provide direct links to other tools and networks (McGown et al. 1998). However, other researchers also claim that as current digital design tools are created to produce precise and structured geometry, they do not support fluid exploration of design alternatives (Prats et al. 2009). This was confirmed by Pipes (2007) in an earlier study who noted that computer-based methods demand precision too soon in the process that steers designers away from design thinking. Other researchers claim that although digital sketching has advantages, non-digital sketch methods illustrate the thought processes in a more spontaneous way while digital tools sometimes limit creativity (Dorta et al. 2008; Verstijnen et al. 1998). Examples of digital representations are shown in Figure 3.17 and 3.18 that describe the external aesthetics of a game controller, as well as the mechanical components of a vehicle.

How sketching has supported creativity is less easy to identify as evidence has only emerged from other fields such as cognitive science, neurophysiology, computing and design research (Goldschmidt 1991a). In an experimental study, Aldoy and Evans (2011) found that graduating students preferred the use of digital methods for renderings and control/Engineering Drawings with 91 percent and 72 percent respectively often/always using digital methods. However, they also found that both graduating students and practitioners felt that the use of digital tools could not entirely replace conventional methods. Coyne et al. (2002) noted important differences between manual and digital skills. Manual skills such as drawing can be applied anywhere. For CAD skills, there is a need to learn different systems and as they evolve, they 'bring the burden of change to the practitioner's consciousness'. This requires the designer to quickly adapt to new technologies or be at risk of being left behind. An ability to utilise CAD tools may rely more

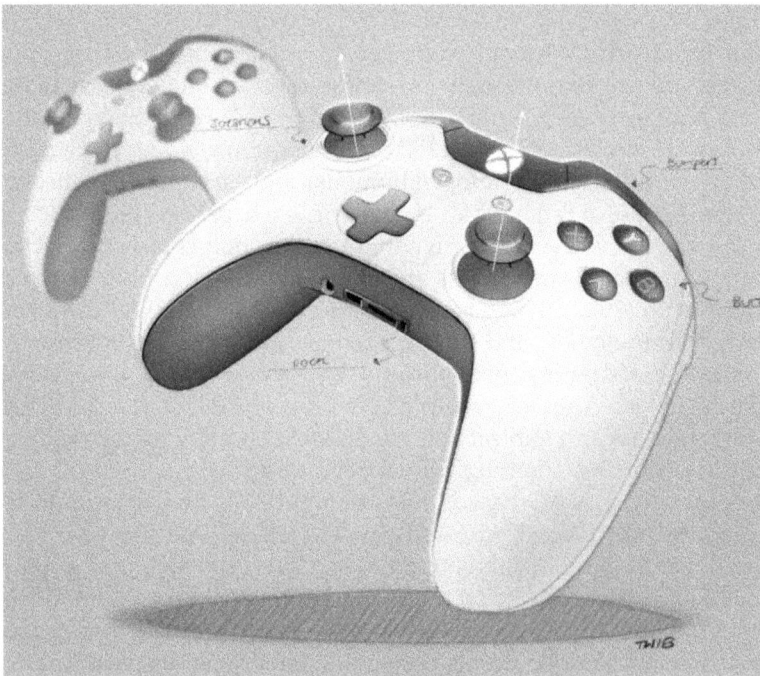

FIGURE 3.17
Presentation Drawings (Walters, n.d.).

FIGURE 3.18
A Technical Illustration (Walters, n.d.).

heavily on the more explicit training of the CAD software and related environment. In contrast, learning to sketch may also require understanding of certain rules and methods (i.e. perspective techniques), but may also require the acquisition of implicit knowledge and skills acquired through opportunities to practice sketch representation.

3.6 Manual 3D media

Manual methods of working on physical media often take the form of small pieces of paper and cardboard to large clay Models. More popular materials include paper, cardboard, plastic sheets, clay, balsa wood and expanded polystyrene foam. Tools include solvent glue, wires, epoxy resin, crafting knives, hot glue guns, files, sandpapers and spray paints. The use of manual 3D media allows a hands-on approach to explore and evaluate the design that may be too complex to visualise on computer. For example, rigid cellular foam can be sculpted by hand quickly as compared to creating a model digitally on the computer. Manual media allow developers to convey the overall form without the need for colour, texture or weight (Garner 2006). To improve realism, parts may be finished to a high level of detail by applying colour, textures, decals and ready-made parts such as buttons and LED lights. Electrical handheld tools such as the Dremel multi-tool can be very useful for drilling, cutting, polishing, sanding, carving and engraving (Figures 3.19 and 3.20).

FIGURE 3.19
An Appearance Model (Dareshani, n.d.).

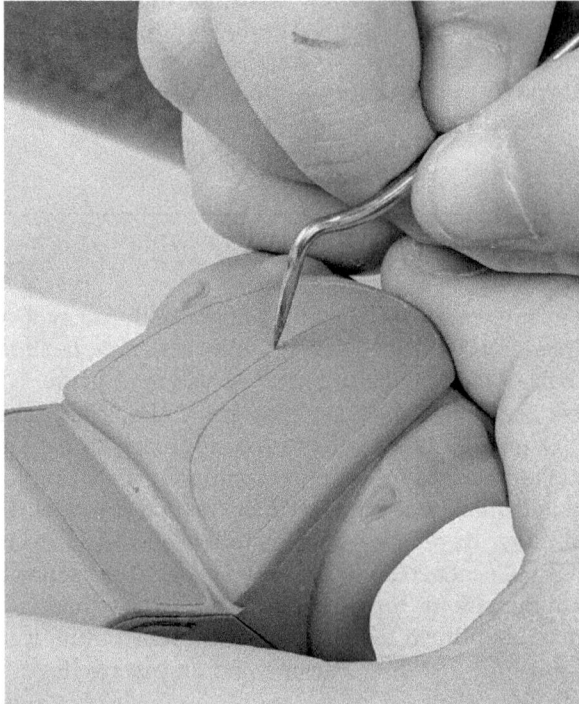

FIGURE 3.20
Working on an Appearance Model (Dareshani, n.d.).

3.7 Digital 3D media

The use of computers sometimes conveys to the designer that they are in a more advanced stage of the design process and the 3D sketches require more detail and better finish as compared to simple conceptual drawings, thereby influencing the way the final result is produced (Bilda and Demirkan 2003; Madrazo 1999; Stones and Cassidy 2010). Knoop et al. (1996) claimed that computer support of later stages of the design process is easier to achieve as the product description has already been defined. Digital 3D media is associated with the use of Computer Aided Design (CAD) that uses computers to produce digital visual design representations that have an illusion of depth. The four main advantages of using CAD include faster speed, greater precision, more efficient modifications and ease of information transfer (Schweikardt and Gross 2000). Other advantages include reproducing the design as a photo realistic image or viewing it from various angles with a choice of lighting, environment, colour or texture. Bermúdez and King (1998) found that digital media such as 3D CAD is often more suited for design development, whereas manual media are more appropriate for ideation stages.

CAD software is used to produce a 2D or 3D Drawing or Model and for performance analysis such as calculations (Figure 3.21) and stress-strain simulations including Finite Element Analysis (FEA) (Figure 3.22). The data can also be used for tooling or for additive manufacturing. A sketching input device that has gained popularity in recent years is the use of a graphic tablet with software such as ZBrush, Mudbox or 3D Coat could allow designers to quickly develop 3D digital mock-ups with close to the same flexibility as working on paper and pencil. This form of digital sculpting utilises polygonal meshes where strokes on the graphic tablet deforms the geometry and material can be added or

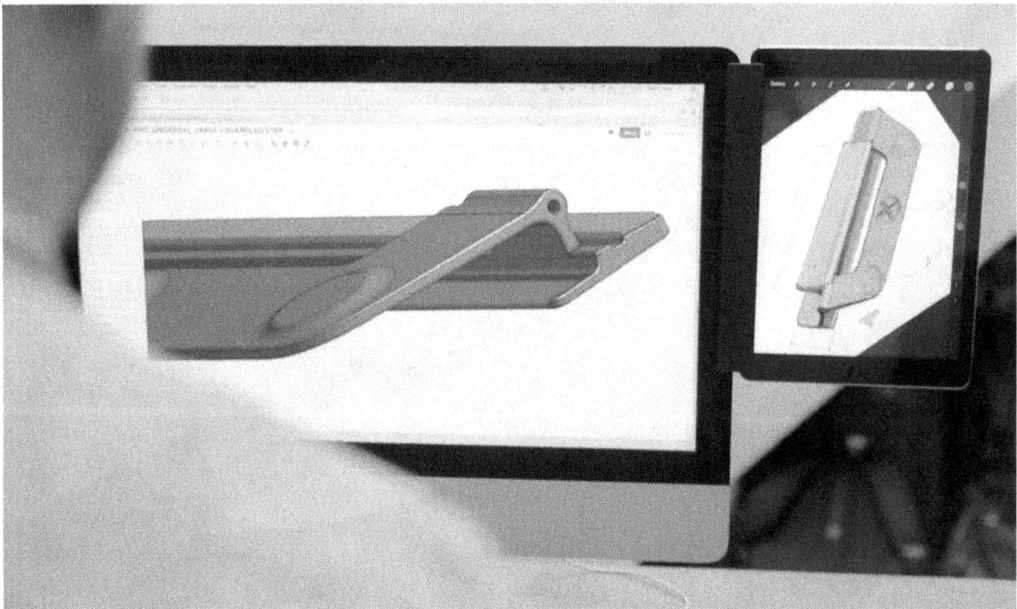

FIGURE 3.21
CAD (Computer Aided Design) modelling (Diaz, n.d.).

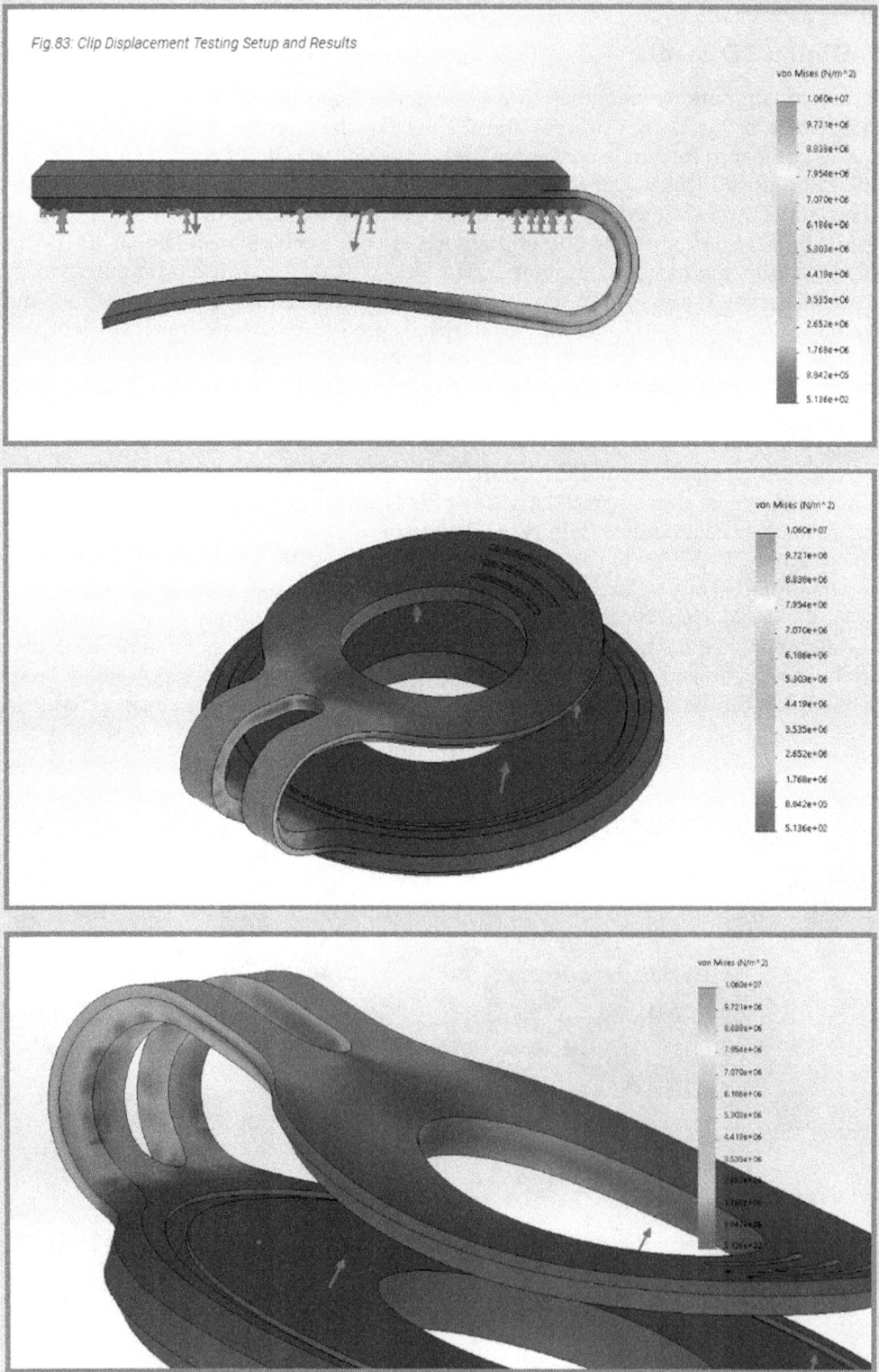

FIGURE 3.22
Finite element analysis using 3D CAD (Choudhury, n.d.)

removed. The use of digital clay using stroke-based input devices may establish a new way of working for designers. Digital sculpting is different from traditional modelling whereby the process is closer to artistic production as compared to CAD methods. The graphic tablet as an input device, recognising different pressure levels and freehand gestures, similar to traditional sketching skills. Van Dijk (1995) proposed a sketch-based modelling system capable of dealing with 2D hand-sketched curves and transforming them into 3D volumes that can be edited. Sequin (2005) suggested a software that could combine CAD with natural hands-on interaction in the form of an interactive digital clay modelling approach. Bae et al. (2008) also proposed a system that could work with 3D strokes and wireframes but without any volume.

Stacey and Eckert (2003) indicated that because 2D or 3D digital media are precise, the perceptual interpretation space for viewers are narrow. CAD reduces ambiguity and this clarity allows for more accurate visualisation of the design intent. The use of *Computer-Aided Industrial Design* (CAID) refers to the use of dedicated software that generates complex organic shapes and rendering them with a high degree of realism. According to Potter (2000), CAID allows for creativity by letting the user experiment with shapes and forms; whilst traditional CAD is more structured and precise to meet engineering needs. To ensure that the original design intent is retained, it is important that the industrial designer should take the lead for CAID modelling. According to Siu and Dilnot (2001), computers and digital tools can vastly reduce the time-to-market by integrating idea conceptualisation, virtual form-giving and form-making processes into a single-phase process.

Digital tools have a relatively common approach to use that allows designers, engineers and managers to communicate more efficiently (Lynn 2006). Marx (2000) suggested that digital methods allow designs to be changed and manipulated easily, compared to traditional methods that require more physical effort. Digital methods provide clients with a clearer understanding through the use of realistic renderings. They also reduce the number of physical Models required in the design process (Catalano et al. 2002). The drawback is that it can sometimes be unproductive for innovation as the familiarity of the user interface within the computer programme can become a distraction for the designer (Stacey and Eckert 2003; Dorta *et al.* 2008). Ibrahim and Rahimian (2010) claim that although current CAD tools can be advantageous for communicating detailed engineering design, the use of CAD can sometimes hinder the creativity of novice designers due to the inherent characteristics of digital tools.

The majority of geometric modelling software requires a high degree of specialisation from users to achieve the forms that they desire. In addition, 3D CAD data may sometimes be misleading. For example, Lueptow (2000) explains that a circular hole could be wrongly viewed as an ellipse in an isometric 3D projection. Campbell *et al.* (2006) also pointed out that the developer may lose his 'hands-on' sense of interacting with a physical model if there is an over-reliance on 3D digital media. Bilda and Demirkan (2003) claimed that current CAD software is not sufficiently flexible to support activities such as doodling, drawing, and copying that are all part of the visual thinking and reasoning processes. Despite these drawbacks, more recent studies have provided evidence that advancement in CAD now supports the conceptual phase of the design process, in which accurate visualisations made possible with CAD modelling can help designers alter and refine their design thinking (Salman et al. 2014). Gesture-based input devices and 3D modelling software with more intuitive user interfaces have also improved, although such an approach may still impose an additional cognitive load on the user (Zhai and Milgram 1993). To counter such constraints, the use of Virtual Reality (VR) has shown to be promising by enabling true 3D interactions (Jin and Lee 2019). Di Gironimo et al. (2006)

and Bruno and Muzzupappa (2010) showed that the VR environment could be useful for Prototype evaluation as the technology has the capability to replicate the actual physical product. Bowman et al. (2004) define VR as a 'human-computer interaction in which user's tasks are performed directly in a 3D spatial context', and the use of VR is direct which reduces the number of actions during manipulation of design objects and reduces working memory load (Lee and Yan 2016).

3.8 Summary

Chapter 3 has discussed the role and use of design representation when applied in design practice. In the Chapter's first section (Section 3.1), we have discussed the purpose of design representation in terms of design thinking in practice. In particular, we have reviewed current works towards understanding how design representation's use in practice may be influenced by the evolving requirements representations are employed to address during a process of design. Our discussion has attempted to ground a definition of design representation through an exploration of representation's use to support the requirements of design practice in terms of thinking towards possible solution candidates, their development and implementation. In this, we have also discussed the role of communication through representation in relation to the practical requirements of design.

The Chapter's opening section has examined existing work towards understanding how the application of representation in practice may relate to particular requirements related to design thinking. However, departing from the discussion of design thinking and representation presented in the previous chapter (Chapter 2), the section considered needs around design thinking in terms of the features and characteristics of representations themselves. Sketching, as ubiquitous media of representation, was discussed with various types of sketch, their role and their use in practice identified. By means of comparison, digital media in particular, CAD (Computer-Aided Design) tools were discussed as means to highlight how the choice of media may implicate representation, cognition and communication of ideas at various phases of the design process.

Section 3.2 introduced existing work towards classification and taxonomy of design representations used in practice. Various approaches to identify and classify design representations were discussed in terms of their use in communication between stakeholders and their role in the evolving requirements of the design process. In particular, more extensive work towards understanding types of sketch representation, their physical characteristics, and purpose of use were discussed.

The Chapter's final sections considered relations between the features of different tools or media of design representation and their use in design practice. Here we have divided design media between digital (i.e. CAD, CAID) and Manual (i.e. hand-sketching) tools, in a comparative discussion of the strengths and limitations of both approaches. In this, we have attempted to provide an initial mapping of the strengths and limitations of different media of expression in relation to their use in design representation. The section concluded with a discussion of various *2D* and *3D Manual* and *Digital* media as examples of how various tools are used.

In this chapter, we have contextualized our discussion of design representation's relationship with design thinking in Chapter 2 through a discussion of representation's application in practice. To achieve this, we have focused upon types of design representation, and how differences in representational type may implicate the various requirements of design practice (i.e. communication between stakeholders, changing needs of the design process). In introducing media of expression as an influence on representation, we highlight how differences in the representation derived from and articulated through media of expression may be dependent upon the representational abilities of various tools. In doing so, we provide the foundations for the following chapter's identification and discussion of a taxonomy of design representations, their relation to design thinking and process.

References

Alisantoso, D. L., Khoo, P., et al. (2006) A Design Representation Scheme for Collaborative Product Development. *International Journal of Advanced Manufacturing Technology*, (30), pp. 30–39.

Aldoy, N. and Evans, M. (2011) A Review of Digital Industrial and Product Design Methods in UK Higher Education. *The Design Journal*, Vol. 14(3), pp. 343–368.

Andreasen, M. M. (1994) Modelling - The Language of the Designer. *Journal of Engineering Design*, Vol. 5(2), pp. 103–115.

Arnheim, R. (1969) *Visual Thinking*. CA: University of California Press.

Brown, A. G. P. (2003) Visualization as a Common Design Language: Connecting Art and Science. *Automation in Construction*, Vol. 12, pp. 703–713.

AskOxford (2008) *Dictionary Resources from Oxford University*. http://www.askoxford.com/ dictionaries/?view=uk. Accessed on 21 October 2008.

Bae, S. H., Balakrishnan, R., et al. (2008) ILoveSketch: As-Natural-as-Possible Sketching System for Creating 3D Curve Models. Quoted in: Cousins, S. B. and Beaudouin-Lafon, M. *Proceedings of the 21st Annual ACM Symposium on User Interface Software and Technology*, pp. 151–160.

Bar-Eli, S. (2013) Sketching Profiles: Awareness to Individual Differences in Sketching as a Means of Enhancing Design Solution Development. *Design Studies*, Vol. 34(4), pp. 472–493.

Baskinger, M. (2008) Pencils Before Pixels: A Primer in Hand-Generated Sketching. *Interactions*. March–April, pp. 28–36.

Baber, C. (2003) *Cognition and Tool Use: Forms of Engagement in Human and Animal Use of Tools* (1st ed.). London: Taylor and Francis.

Bermúdez, J. and King, K. (1998) Media Interaction & Design Process: Establishing a Knowledge Base.Proceedings of the ACADIA Conference. Québec City, Canada. October 22-25, pp. 6–25.

Bilda, Z., Gero, J. S., et al. (2006) To Sketch or Not to Sketch? That is the Question. *Design Studies*, Vol. 27(5), pp. 587–613.

Bilda, Z. and Demirkan, H. (2003) An Insight on Designers' Sketching Activities in Traditional Versus Digital Media. *Design Studies*, Vol. 24(1), pp. 27–50.

Bly, S. A. (1988a) A Use of Drawing Surfaces in Different Collaborative Settings. *Proceedings of Computer Supported Cooperative Work '88*, Portland, OR, ACM Press, New York.

Boote, S. (2006) Visualization - Visible Benefits. *New Design*, Vol. 35, p. 44.

Bowman, D. A., Kruijff, E., et al. (2004) *3D User Interfaces: Theory and Practice*. London: Addison Wesley.

Brandt, E. (2005) How do Tangible Mock-ups Support Design Collaboration? *Proceedings of theNordic Design Research Conference*, 'In the Making', Copenhagen, Denmark.

Bruno, F. and Muzzupappa, M. (2010) Product Interface Design: A Participatory Approach Based on Virtual Reality. *International Journal of Human-Computer Studies*, Vol. 68(5), pp. 254–269.

Buur, J. and Andreasen, M. M. (1989a) Design models in mechatronic product development. *Design Studies*, Vol. 10(3), pp. 155–162.

Buxton, B. (2007) *Sketching User Experiences - Getting the design right and the right design*. San Francisco: Morgan Kaufmann.

Cain, R. E. (2005) *Involving users in the Design Process: The Role of Product Representations in Co-Designing* (PhD Thesis). Department of Design and Technology. Loughborough: Loughborough University.

Campbell, R. I., Hague, R. J., et al. (2006) The Potential for the Bespoke Industrial Designer. *The Design Journal*, Vol. 6(3), pp. 24–26.

Cardella, M. E., Atman, C. J. et al. (2006) Mapping Between Design Activities and External Representations for Engineering Student Designers. *Design Studies*, Vol. 27(1), pp. 5–24.

Catalano, C. E., Falcidieno, B., et al. (2002) A Survey of Computer-Aided Modeling Tools for Aesthetic Design. *Journal of Computing and Information Science in Engineering*. Vol. 2, pp. 11–20.

Chen, K. and Owen, C. L. (1998) A Study of Computer-Supported Formal Design. *Design Studies*, Vol. 19, pp. 331–359.

Chiu, M. -L. (2002) An Organisational View of Design Communication in Design Collaboration. *Design Studies*, Vol. 23(2), pp.187–210.

Cook, M. T. and Agah, A. (2009) A Survey of Sketch-Based 3-D Modelling Techniques. *Interacting with Computers*, Vol. 21(3), pp. 201–211.

Cooper, D. (2018) Imagination's Hand: The Role of Gesture in Design Drawing. *Design Studies*, Vol. 54, pp. 120–139.

Coyne, R., Park, H., et al. (2002) Design Devices: Digital Drawing and the Pursuit of Difference. *Design Studies*, Vol. 23(3), pp. 263–286.

Cross, N. (2007) *Designerly Ways of Knowing*. Basel, Switzerland: Birkhauser Press.

Cross, N. (2008) *Engineering Design Methods: Strategies for Product Design* (4th ed.). Chichester: John Wiley and Sons.

Di Gironimo, G., Lanzotti, A., et al. (2006) Concept Design for Quality in Virtual Environment. Computers and Graphics, 30(6), pp. 1011–1019.

Do, E. Y.-L. (2005) Design Sketches and Sketch Design Tools. *Knowledge-Based Systems*,(18), pp. 383–405.

Do, E. Y-L. (2002) Drawing Marks, Acts, and Reacts: Toward a Computational Sketching Interface for Architectural Design. *Artificial Intelligence for Engineering Design, Analysis and Manufacturing*, Vol. 16(3), pp. 149–171.

Dorish, P. (2001) *Where the Action is: The Foundations of Embodied Interaction*. London: MIT Press.

Dorta, T. Perez, E., et al. (2008) The Ideation Gap: Hybrid Tools, Design Flow and Practice. *Design Studies*, 29(2), pp. 121–141.

Dorta, T. (2008) *Design Flow and Ideation*. http://www.din.umontreal.ca/documents/dorta/22-IJAC.pdf. Accessed on 12 October 2008.

Dorta, T. and Pérez, E. (2006) *Hybrid Modeling: Revaluing Manual Action for 3D Modeling*. http://www.din.umontreal.ca/documents/dorta/12-ACADIA2006b.pdf. Accessed on 15 October 2008.

Dorta, T. (2005) *Hybrid Modeling: Manual and Digital Media in the First Steps of the Design Process*. http://www.din.umontreal.ca/documents/dorta/9-eCAADe2005.pdf. Accessed on 12 October 2008.

Eckert, C. and Boujut, J.-F. (2003) The Role of Objects in Design Co-Operation: Communication through Physical or Virtual Objects. *Computer Supported Cooperative Work*, Vol. 12(2), pp. 145–151.

Eckert, C. M. and Stacey, M. K. (2000) Sources of Inspiration: A Language of Design. *Design Studies*, Vol. 21(5), pp. 523–538.

Engelbrektsson, P. and Soderman, M. (2004) The Use and Perception of Methods and Product Representations in Product Development: A Survey of Swedish Industry. *Journal of Engineering Design*, Vol. 15(2), pp. 141–154.

Evans, M., Wallace, D., et al. (2004) An Evaluation of Haptic Feedback Modelling During Industrial Design Practice. *Design Studies*, Vol. 26(5), pp. 487–508.

Evans, M. A. (2002) *The Integration of Rapid Prototyping within Industrial Design Practice* (Staff Thesis). Department of Design and Technology. Loughborough: Loughborough University.

Ferguson, E. S. (1992) *Engineering and the Mind's Eye*. Cambridge, Massachusetts: MIT Press.

Fish, J. (2004) Cognitive Catalysis: Sketches for a Time-lagged Brain. In: Goldschmidt, G. and Porter, W.(eds.), *Design Representation*. London: Springer, pp. 151–184.

Fish, J. C. (1996) *How Sketches Work-A Cognitive Theory for Improved System Design* (PhD dissertation). Loughborough University of Technology. Quoted in: Do, E., Gross, M., D., Neiman, B. and Zimring, C. (2000) Intentions in and relations among design drawings. *Design Studies* Vol. 21(5).

Gantz, C. M. (2005) What a Difference 50 Years Makes!. *Innovation: The Quarterly Journal of the Industrial Designers Society of America*, Vol. 24(1), pp. 20–21.

Garner, S. (2006) *Modelling Workbook 1: T211 Design and Designing Workbook 1 Technology Level 2* (2nd ed.). Milton Keynes: The Open University.

Garner, S. (2004) *T211 Design and Designing - An Introduction to Design and Designing*. Milton Keynes: The Open University Press.

Goldschmidt, G. (2007) To See Eye to Eye: The Role of Visual Representations in Building Shared Mental Models in Design Teams. *CoDesign*, Vol. 3(1), pp. 43–50.

Goldschmidt, G. and Porter, W. (2004) *Design Representation*. London: Springer-Verlag London Ltd.

Goldschmidt, G. (1994) On Visual Design Thinking: The vis Kids of Architecture. *Design Studies* Vol. 15(2), pp. 158–174. Quoted in: Purcell, A., T. and Gero, J., S. (1998) Drawings and the Design Process. *Design Studies*, Vol. 19(4).

Goldschmidt, G. (1991a) The Dialectics of Sketching. *Creativity Research Journal*, Vol. 4(2), pp. 123–143.

Goldschmidt, G. (1989) Problem Representation Versus Domain of Solution in Architectural Design Teaching. *Journal of Architectural and Planning Research*, Vol. 6(3), pp. 204–215. Quoted in: Purcell, A. T. and Gero, J. S. (1998) Drawings and the Design Process. *Design Studies*, Vol. 19(4).

Gorner, R. (1994) Zur psychologischen Anayse von Konsteukteur - und Entwurfstatigkeiten. In: Die Handlungsregulationstheorie: Von der Praxis einer Theorie, Bergmann, B. and Ritchter, P. (eds.). Gottingen: Hogrege, pp. 233–241. Quoted in: Goldschmidt, Gabriela and Porter, W., L. (2004) *Design Representation*. London, Springer-Verlag.

Gutwin, C. and S. Greenberg (2002) A Descriptive Framework of Workspace Awareness for Real-Time Groupware. *Computer Supported Cooperative Work*, Vol. 11, (3), pp. 411–446.

Heath, C. and P. Luff (1991) Collaborative Activity and Technological Design: Task Coordination in London Underground Control Rooms. *Proceedings of ECSCW '91*, Amsterdam, pp. 65–80. Quoted in: Perry, M. and Sanderson, D. (1998) Coordinating Joint Design Work: The Role of Communication and Artifacts. *Design Studies*, Vol. 19(3).

Hendry, D. G. (2004) Communication Functions and the Adaptation of Design Representations in Interdisciplinary Teams. Symposium on Designing Interactive Systems: *Proceedings of 2004 Conference on Designing Interactive Systems Processes Practices Methods and Techniques*.

Holmquist, L. E. (2005) Prototyping: Generating Ideas or Cargo Cult Designs? *Interactions*, Vol. 12(2), pp. 48–54.

Ibrahim, R. and Rahimian, F. P. (2010) Comparison of CAD and Manual Sketching Tools for Teaching Architectural Design. *Automation in Construction*, Vol. 19(8), pp. 978–987.

Jin, Y. and Lee, S. (2019) Designing in Virtual Reality: A Comparison of Problem-Solving Styles Between Desktop and VR Eenvironments. *Digital Creativity*, Vol. 30(2), pp. 107–126.

Johansson, P., Persson, S., et al. (2001) The Use of Product Representations in Industry - A Survey Dealing with Product Development in Sweden. *Proceedings ofSKFC'01, The 3rd International Workshop on Strategic Knowledge and Concept Formation*. Sydney, Australia. Quoted in: Persson, S. (2002) *Industrial Design – Engineering Design Interaction – Studies of Influencing Factors in Swedish Product Developing Industry*. Thesis for the degree of Licentiate of Engineering, Chalmers University of Technology, Göteborg, Sweden.

Jones, C. J. (1992) *Design Methods* (2nd ed.). New York: John Wiley and Sons.

Jonson, B. (2005) Design Ideation: The Conceptual Sketch in the Digital Age. *Design Studies*, Vol. 26(6), pp. 613–624.

Johnson, S. (1998) What's in a Representation, Why do We Care, and What does it Mean? Examining Evidence from Psychology. *Automation in Construction*, (8), pp. 15–24.

Kalay, Y. (2004) *Architecture's New Media*. Cambridge: MIT Press.

Kaplan, R. and Kaplan, S. (1982) *Cognition and Environment: Functioning in an Uncertain World*. New York: Praeger Publishers.

Knoop, W. G., Breemen, E. J. J., et al. (1996) Towards More Effective Capturing of Empirical Data from Design Processes. *Proceedings of 1st Conference in Descriptive Design*, Istanbul.

Koutamanis, A. (1993) The Future of Visual Design Representations in Architecture. *Automation in Construction*, Vol. 2(1), pp. 47–56.

Kurtoglu, T. and Stahovich, T. F. (2002) Interpreting Schematic Sketches using Physical Reasoning. *In AAAI 2002 Spring Symposium Series, Sketch Understanding*.

Lakin, F. (1990) Visual Languages for Cooperation: A Performing Medium Approach to Systems for Cooperative Work. In: Gategher, J., Kraut, R. E. and Egldo, C. (eds.) (1990) Intellectual/ Teamwork: Social and technological Foundations of Cooperative Work. *Lawrence Edbaum Associates*, pp. 453–488. Quoted in: Peng, C. (1994) Exploring Communication in Collaborative Design: Co-operative Architectural Modelling. *Design Studies*, Vol. 15(1).

Lawson, B. (2004) *What Designers Know*. Oxford: Architectural Press.

Lawson, B. (1997) *How Designers Think – the Design Process Demystified*. Oxford: Architectural Press. Quoted in: Persson, S. and Warell, A. (2003) Relational Modes between Industrial Design and Engineering Design – a Conceptual Model for Interdisciplinary Design Work. *Proceedings of the 6th Asian Design International Conference*, Tsukuba.

Lawson, B. (1994) *Design in Mind*. Oxford: Butterworth Architecture. Quoted in: McGown, A.; Green, G. and Rodgers, P., A. (1998) Visible Ideas: Information Patterns of Conceptual Sketch Activity. *Design Studies*, Vol. 19(4), pp. 431–453.

Lee, S. and Yan, J. (2016) The Impact of 3D CAD Interfaces on User Ideation: A Comparative Analysis Using Sketchup and Silhouette Modeler. *Design Studies*, Vol. 44, pp. 52–73.

Leonard-Barton, D. (1991) Inanimate Integrators: A Block of Wood Speaks. *Design Management Journal*, Summer 1991 pp. 61–67. Quoted in: Söderman, M. (2002) Comparing Desktop Virtual Reality with Handmade Sketches and Real Products: Exploring Key Aspects for End-Users' Understanding of Proposed Products. *Journal of Design Research*, Vol. 2(1).

Lipson, H. and Shpitalni, M. (2000) Conceptual Design and Analysis by Sketching. *Artificial Intelligence for Engineering Design, Analysis and Manufacturing*, Vol. 14(5), pp. 391–401.

Logan, G. D. and Radcliffe, D. F. (2000) Videoconferencing to Support Designing at a Distance. Quoted in: Scrivener, S., Ball, A., R., Linden J. and Woodcock, A. (2000) Collaborative Design. *Proceedings of CoDesigning 2000*. London: Springer- Verlag, pp. 364.

Longman Dictionary of Contemporary English (2005) Essex: Pearson Education Limited.

Löwgren, J. (2004) Animated Use Sketches as Design Representations. *Interactions*, Vol. 11(6), pp. 22–27.

Lueptow, R. M. (2000) *Graphics Concepts*. New Jersey: Prentice Hall.

Lynn, D. (2006) Automotive Design Education Embraces the Digital Age. In: Cullen, C. (ed.), *Eastman IDSA National Education Symposium Proceedings*, Austin, Texas. Dulles, VA: *Industrial Designers Society of America*, pp. 107–114.

Madrazo, L. (1999) Types and Instances: A Paradigm for Teaching Design with Computers. *Design Studies*, Vol. 20(4), pp. 177–193.

Marx, J. (2000) A Proposal for Alternative Methods for Teaching Digital Design. *Automation in Construction*, Vol. 9, pp. 19–35.

McGown, A., Green, G., et al. (1998) Visible Ideas: Information Patterns of Conceptual Sketch Activity. *Design Studies*, Vol. 19(4), pp. 431–453.

McKoy, F. L., Vargas-Hernández, N., et al. (2001) Influence of Design Representation on Effectiveness of Idea Generation. *Proceedings ofDETC'01: ASME* 2001 Design Engineering Technical Conferences and Computers and Information in Engineering Conference, Pittsburgh, Pennsylvania, September 9–12, 2001.

Olofsson, E. and K. Sjölén (2005) *Design Sketching*. Sundsvall, Sweden: Keeos Design Books AB.

Palmer, S. E. (1987) Fundamental Aspects of Cognitive Representation. In: Roch, E. and Lloyds, B. B. (eds.) (1987) *Cognition and Categorization*. Hillsdale, NJ: Lawrence.

Pavel, N. (2005) *The Industrial Designer's Guide to Sketching*. Trondheim: Tapir Academic Press.

Perry, M. and Sanderson, D. (1998) Coordinating Joint Design Work: The Role of Communication and Artifacts. *Design Studies*, Vol. 19(3), pp. 273–288

Pipes, A. (2007) *Drawing for Designers*. London: Laurence King Publishing.

Potter, N. (2002) *What is a Designer: Things.Places.Messages* (4th ed.). London: Hyphen Press.

Potter, C. (2000) The CAID Connection. *Computer Graphics World*, Vol. 23(3), pp. 21–28. In: Evans, M., A. (2002) *The Integration of Rapid Prototyping within Industrial Design Practice* (Staff Thesis). Loughborough: Department of Design and Technology Loughborough University.

Prats, M., Lim, S., Jowers, I., Garner, S. and Chase, S. (2009) Transforming Shape in Design: Observations from Studies of Sketching. *Design Studies*, Vol. 30(5), pp. 503–520.

Purcell, T. and Gero, J. (1999) Design and Other Types of Fixation. *Design Studies*, Vol. 17(4), pp. 363–383.

Purcell, A. T. and Gero, J. S. (1998) Drawings and the Design Process. *Design Studies*, Vol. 19(4), 389–430.

Rahimian, R. F., Ibrahim, R., et al. (2008) Feasibility Study on Developing 3D Sketching in Virtual Reality (VR) Environment. ALAM CIPTA. *International Journal of Sustainable Tropical Design Research and Practice*, Vol. 3, pp. 60–78.

Robertson, T. (1996) Embodied Actions in Time and Place: The Cooperative Design of a Multimedia Educational Computer Game. *CSCW*, Vol. 5(4), pp. 341–367. Quoted in: Perry, M. and Sanderson, D. (1998) Coordinating Joint Design Work: The Role of Communication and Artifacts. *Design Studies*, Vol. 19(3).

Romer, A., Pache, M., et al. (2001) Effort-Saving Product Representations in Design Results of a Questionnaire Survey. *Design Studies*, Vol. 22(6), pp. 473–491.

Saddler, H. J. (2001) Understanding Design Representations. *Interactions*, Vol. 8(4), pp. 17–24.

Salman, H. S., Laing, R., and Conniff, A. (2014) The Impact of Computer Aided Architectural Design Programs on Conceptual Design in Educational Context. *Design Studies*, Vol. 35(4), pp. 412–439.

Schmidt, K. and Wagner, I. (2004) Ordering Systems: Coordinative Practices and Artifacts in Architectural Design and Planning. *Computer Supported Cooperative Work*, Vol. 13, pp. 349–408.

Schön, D. and Wiggins, G. (1992) Kinds of Seeing and their Functions in Designing. *Design Studies*, Vol. 13(2), pp. 135–156.

Schön, D. (1983) *The Reflective Practitioner: How Professionals think in Action*. London: Temple Smith.

Schrage, M. (1993) The Cultures of Prototyping. *Design Management Journal*, Winter, pp. 55–65.

Schweikardt, E. and Gross, M. (2000) Digital Clay: Deriving Digital Models from Freehand Sketches. *Automation in Construction*, Vol. 9(1), 107–115.

Scrivener, S. A. R. and Clark, S. M. (1994) Sketching in collaborative design. In MacDonald, L. and Vince, J. (eds.) *Interacting with Virtual Environments*. Chichester: Wiley. Quoted in: Van der Lugt, R. (2005) How Sketching can Affect the Idea Generation Process in Design Group Meetings. *Design Studies*, Vol. 26(2).

Seitamaa-Hakkarainen, P. and Hakkarainen, K. (2000) Visualization and Sketching in the Design Process. *Design Journal*, Vol. 3(1), pp. 3–14.

Self, J., Evans, M., et al. (2014) The Influence of Expertise upon the Designer's Approach to Studio Practice and Tool Use. *The Design Journal*, 17(2), pp. 169–193.

Self, J (2012) Sketching vs. CAD, Why Ask?. *Core77 Design Periodical*. New York: Core77.

Sequin, C. H. (2005) CAD Tools for Aesthetic Engineering. *Computer-Aided Design*, Vol. 37(7), pp. 737–750.

Shih, Y. T., Sher, W. D., et al. (2017) Using Suitable Design Media Appropriately: Understanding How Designers Interact with Sketching and CAD Modelling in Design Processes. *Design Studies*, Vol. 53, pp. 47–77.

Siu, N. W. C. and Dilnot, C. (2001) The Challenge of the Codification of Tacit Knowledge in Designing and Making: A Case Study of CAD Systems in the Hong Kong Jewellery Industry. *Automation in Construction*, Vol. 10, pp. 710–714.

Song, S. and Agogino, W. D. (2004) Insights on designers' sketching activities in product design teams. In: *ASME Design Engineering Technical Conference '04* held in Salt Lake City, Utah. Quoted in: Yang, M., C. and Daniel, J. (2005) A Study of Prototypes, Design Activity, and Design Outcome. *Design Studies*, Vol. 26(6).

Stacey, M. and Eckert, C. (2003) Against Ambiguity. *Computer Supported Cooperative Work*, Vol. 12, pp. 153–183.

Stolterman, E., McAtee, J., et al. (2008) Designerly Tools. *Proceedings of DRS2008, Design Research Society Biennial Conference*, Sheffield, UK, 16–19 July 2008, Sheffield University.

Stones, C., and Cassidy, T. (2010) Seeing and Discovering: How do Student Designers Reinterpret Sketches and Digital Marks During Graphic Design Ideation? *Design Studies*, Vol. 31(5), pp. 439–460.

Suri, J. F. (2003) The Experience of Evolution: Developments in Design Practice. *Design Journal*, Vol. 6(2), pp. 39–48.

Suwa, M. and Tversky, B. (1997) What do Architects and Students Perceive in their Design Sketches? A Protocol Analysis. *Design Studies*, Vol. 18(4), pp. 385–403.

Suwa, M. and Tversky, B. (1996) What Architects See in Their Sketches: A Protocol Analysis. *Artificial Intelligence in Design '96*. Stanford University.

Synder, C. (2002) *Paper Prototyping*. San Francisco, CA, United States: Morgan Kaufmann Publishers. pp. 20–48.

Tang, J. (1989) *Listing, Drawing, and Gesturing in Design: A Study of the Use of Shared Workspaces by Design Teams* (Ph.D. thesis). Stanford University, Stanford, CA. Quoted in: Eckert, C. and Boujut, J. (2003) The Role of Objects in Design Co- Operation: Communication through Physical or Virtual Objects. *Computer Supported Cooperative Work*, (12), pp. 145–151.

Tjalve, E. (1979) *A Short Course in Industrial Design*. London: Butterworth and Co.

Tohidi, M., Buxton, W., et al. (2006) User Sketches: A Quick, Inexpensive, and Effective Way to Elicit More Reflective User Feedback. *Proceedings ofNordiCHI, The Nordic Conference on Human-Computer Interaction*, pp. 105–114.

Tovey, M., Porter, S., et al. (2003) Sketching, Concept Development and Automotive Design. *Design Studies*, Vol. 24(2), pp. 135–153.

Tovey, M (1997) Styling and Design: Intuition and Analysis in Industrial Design. *Design Studies*, Vol 18(1), pp. 5–31.

Tovey, M. (1989) Drawing and CAD in Industrial Design, *Design Studies*, Vol. 10(1), pp. 24–39.

Ullman, D. G., Wood, S. et al. (1990) The Importance of Drawing in the Mechanical Design Process. *Computer and Graphics*, Vol. 14(2).

Ulrich, K. T. and Eppinger S. D. (2003) *Product Design and Development* (3rd ed.). New York: McGraw-Hill.

Ulusoy, Z. (1999) To Design Versus to Understand Design: The Role of Graphic Representations and Verbal Expressions. *Design Studies*, Vol. 20(2), pp. 123–130.

Utterback, J. and Vedin, B. -A., et al. (2006) *Design Inspired Innovation*. Singapore: World Scientific Publishing.

Van Dijk, C. G. C. (1995) New Insights in Computer-Aided Conceptual Design. *Design Studies*, Vol. 16(1), pp. 62–80.

Van Eck, D. (2015) Dissolving the 'problem of the absent artifact': Design Representations as Means for Counterfactual Understanding and Knowledge Generalisation. *Design Studies*, Vol. 39, pp. 1–18.

Van Sommers, P. (1984) *Drawing and Cognition: Descriptive and Experimental Studies of Graphic Production Processes*. Cambridge: Cambridge University Press.

Van Welie, M. and Van der Veer, G. C. (2000) Structured Methods and Creativity: A Happy Dutch Marriage". Quoted in: Scrivener, S., Ball, A., R., Linden J. and Woodcock, A. (2000) Collaborative Design. *Proceedings of CoDesigning 2000*. London: Springer-Verlag, p. 117.

Verstijnen, I. M., van Leeuwen, C., et al. (1998) Sketching and Creative Discovery. *Design Studies*, Vol. 19(4), pp. 519–546.

Visser, W. (2007) Collaborative Designers' Different Representations. *International Conference on Engineering Design (ICED '07)* held in Cite Des Sciences Et De L'Industrie, Paris, France 28–31 August 2007.

Woodbury, R. F. and Burrow, A. L. (2006) Whither Design Space' Artificial Intelligence for Engineering Design. *Analysis and Manufacture*, Vol. 29(2), pp. 63–82.

Wong, Y. Y. (1992) Rough and Ready Prototypes: Lessons from Graphic Design – Posters and Short Talks. *Proceedings of Human Factors in Computing Systems*, New York: ACM Press.

Zhai, S., and Milgram, P. (1993) Human Performance Evaluation of Manipulation Schemes in Virtual Environments. *Paper presented at the IEEE Virtual Reality Annual International Symposium*, Seattle, September 18–22.

4

Sketches

4.1 Sketch representation

A *Sketch* is a preliminary, visual design representation of something without detail for the basis for a more finished product (Dictionary of Art Terms 2003). In this sense, Design representation as Sketches are often ambiguous or at least provide an opportunity for the ambiguity of expression. Sketches may also be less ambiguous depending upon the type of Sketch and its reason for use in practice. Likewise, Sketches may express design intent at higher or lower levels of fidelity. It is this flexibility in representation that has made the designer's Sketch synonymous with design.

Sketches are usually rapidly executed to present only the key elements of an idea. According to Pipes (2007), a Sketch is a collection of visual cues that forms a stylised skin over a product's components. They comprise of informal freehand marks without any use of instruments (Tjalve et al. 1979b) and consist of draft lines, text, dimensions and basic calculations to explain the meaning, context and scale of the design (Ullman et al. 1990; McGown et al. 1998; Stacey and Eckert 2003). In addition, Sketches often include the use of varying line weight to suggest depth, or have over-traced, re-drawn or hatching lines to define a selection and to draw attention to an area (Do 2005; Ling 2006b). Figures 4.1 and 4.2 show the use of different line weights to emphasize different details of the Render as well as the use of white lines to represent shading or reflective materials and reflection.

In terms of visual detail, Tovey et al. (2003) classified five levels of Sketches, similar to that proposed by Chen et al. (2003). The first level consists of uniform monochrome lines with no shading. At a second level, monochrome lines with varied thickness are used with text annotations. At the third level, Sketches incorporate shading. The next level uses shading in colour; while the last level of Sketches encompass colour, shading, shadows, text and dimensions. A study carried out by Do et al. (2000) aimed to determine whether marks made by an architect during design correlate with the type of task. The pilot study with 62 architecture students showed that the designers could understand each other's conventions in Drawings, and practising architects employ similar conventions when designing. For example, they drew bubble diagrams and partitioning lines when working on spatial arrangements; sun symbols and light rays when addressing lighting concerns; and indicate numbers when detailing sizes and dimensions.

Buxton (2007) identified the key characteristics of Sketches as quick, timely, inexpensive, disposable, plentiful and ambiguous. Prats et al. (2009) found that during the explorative stages of the design process, designers are tuned to look for sub-shapes. Recognising, repeating and modifying these shapes provides an opportunity for further

DOI: 10.1201/9781003227694-4

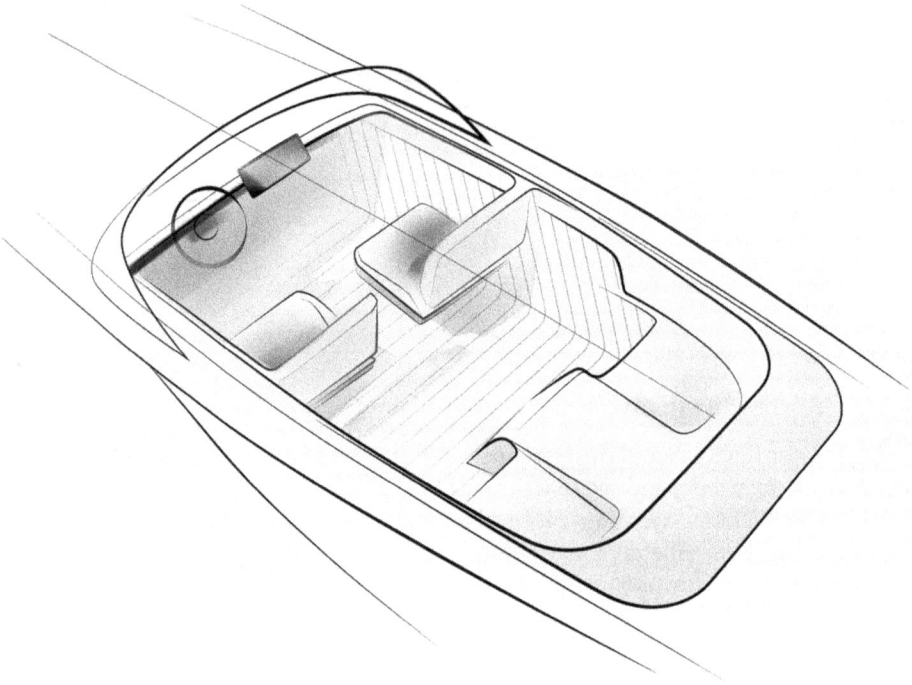

FIGURE 4.1
A digital Rendering (Prototypum, n.d.).

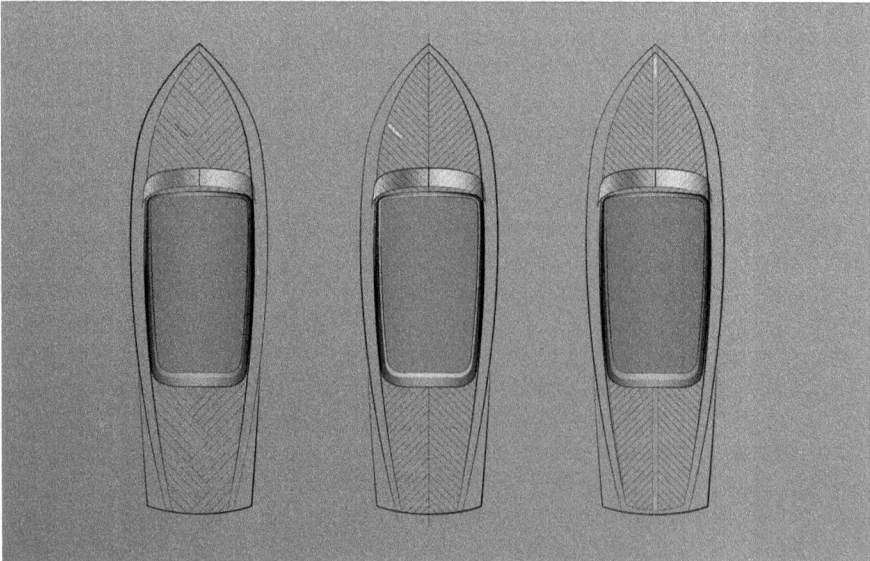

FIGURE 4.2
Single-View Drawings (Prototypum, n.d.).

creative form transformation and development, largely because shapes and sub-shapes allow interpretations in many different ways. During the sketching process, designers are able to focus on different levels of abstraction where local details influence the general structure of the product, and likewise, changes to the structure may influence design details (Garner 2006). On the other hand, engineering designers do not use Sketches to express an idea with realism, but as a means to solve mechanical and production engineering details, as well as as a means to generate solutions (Tovey 1989; Yang and Cham 2007). In contrast, industrial designers use Sketches to represent visual thoughts for communication and reflection to provide opportunity to assess ideas that have been generated (Rodriguez 1992; Ehrlenspiel and Dylla 1993; Fish 1996). Sam Gwilt's rendering in Figure 4.3 shows the careful and deliberate use of shading and different line weights to illustrate the depth and form of the product being communicated.

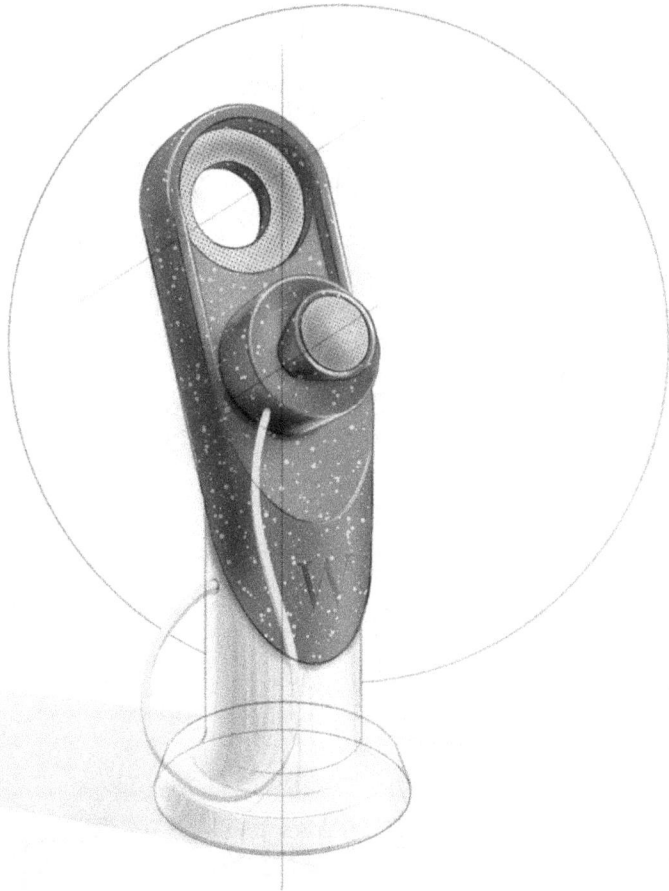

FIGURE 4.3
A product Rendering (Gwilt, n.d.).

Other researchers (Ullman et al. 1990; Ferguson 1992; Van der Lugt 2005) have classified different types of Sketches through their purpose of use in design: *Thinking Sketches* for problem-solving; *Prescriptive Sketches* for providing instructions; *Talking Sketches* for discussion; and *Storing Sketches* to retain ideas. Similarly, Olofsson and Sjölén (2005) grouped *Investigative Sketches* for problem definition; *Explorative Sketches* used for concept generating and evaluating solution ideas; *Explanatory Sketches* to describe and communicate the design; and *Persuasive Sketches* for selling an idea to stakeholders and/or external parties and investors. Among the quickest types of sketch representations are Freehand Sketches for personal use or to aid discussion (Verstijnen et al. 1998). Prats et al. (2009) consider that in creative design, Free-hand Sketches are often utilised for recording ideas that can be useful later in the process, as well as to assist designers in exploring design alternatives such as form and shape in a low-cost, fast and flexible way. Freehand Sketches allow designers to see patterns, discover associations or identify opportunities to inspire further sketching where the reinterpreted forms emerge as new concepts (ibid). This is because Freehand Sketches often provide opportunity for interpretation derived from ambiguity in the expression of design intent. This also relates to a requirement in conceptual design to explore the potential of various solution candidates, before settling on a design direction to pursue in greater detail.

Opportunities for Ideation Sketches to support the exploration of ideas also indicates how the external representation of design intent through freehand sketching acts as a type of cognitive distribution to support design thinking. Freehand sketching provides necessary opportunity for reflection, which is required for a propositional expression to aid evaluation of design ideas as shown in Figure 4.4. As well as Sketches, and often alongside a visual representation of form, colour and materials, other schematic illustrations are made up of symbols, lines, boxes and arrows to denote hierarchical information; while scenarios and storyboards represent the interaction or activities between the user, a product or services.

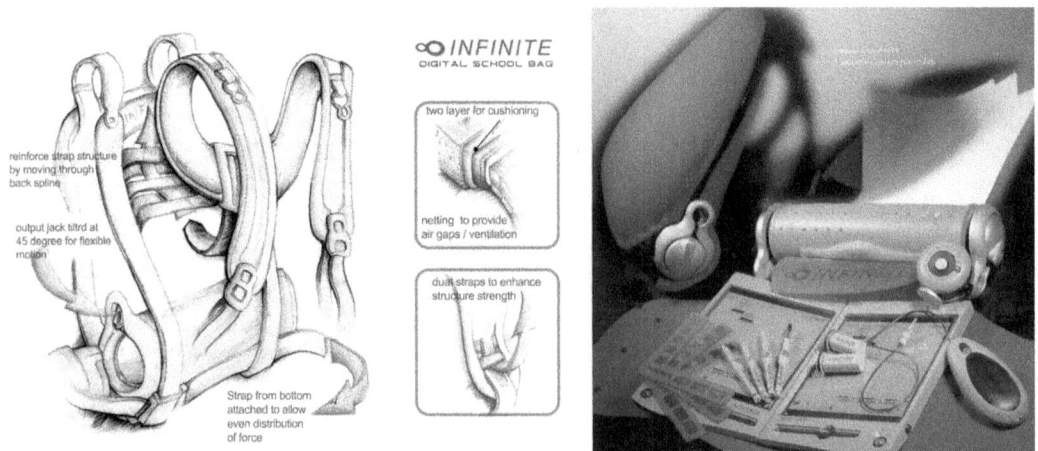

FIGURE 4.4
An Information Sketch next to a product photograph (Chan, n.d.).

4.2 *Idea Sketch*

Idea Sketches are often used in the early design stages for the externalisation, visualisation, exploration and development of ideas (Kojima et al. 1991; Raudebaugh and Newcomer 1999). *Idea Sketches* consist of basic shapes with simple labels and arrows to show the relationship between objects (Moyer 2007). The purpose is to record the idea quickly and to allow the developer to explore other possibilities. *Idea Sketches* are small, ambiguous and require few materials to start with. An *Idea Sketch* is characterised by high degrees of ambiguity to provide space for reflective interpretation of meaning during conceptual design. Likewise, they are low in fidelity, often requiring increased interpretation in the construction of meaning.

Idea Sketches are also known as Thumbnail Sketches (Olofsson and Sjölén 2005), Memo Sketches (Pavel 2005) or Napkin Sketches (Ling 2006a; Baskinger 2008). *Idea Sketches* are 2D visual design representations used at a personal level for externalising thoughts quickly and to show how the design looks as a physical object. They are personal sketches produced to support self-reflection upon the potential of design intent. As they are used during conceptual design ideation, they are often more ambiguous in their expression of design intent. This provides a required level of interpretation as ideas are used to support an exploration of various possibilities, (Figures 4.5 and 4.6).

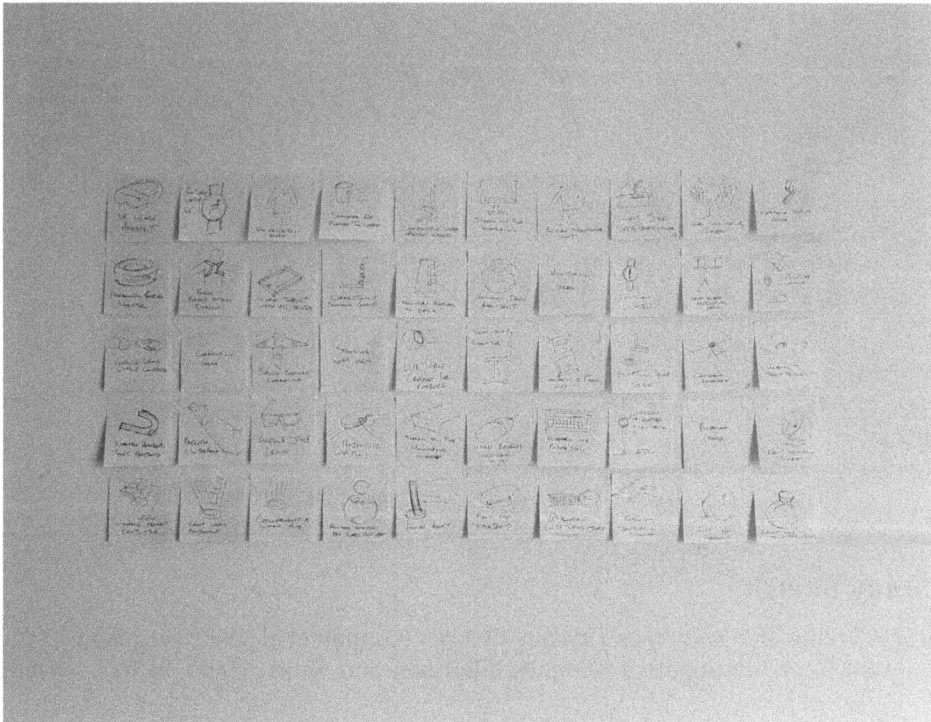

FIGURE 4.5
Idea Sketches generated on post-stick notes (Choudhury, n.d.).

FIGURE 4.6
Idea sketches shown in a sketchbook (Choudhury, n.d.).

4.3 Study Sketch

Study Sketches are also known as *Thinking Sketches* (Ullman et al. 1990; Ferguson 1992; Van der Lugt 2005) or Investigative Sketches (Olofsson and Sjölén 2005). In this sense, they have much in common with reflective-practice (Schön, 1983). Through reflection upon representation through *Study Sketches*, they assist the developer to focus and guide their thoughts about the design's potential and possibilities towards further concept development (Ferguson 1992).

Study Sketches contain few design elements to allow the designer to attempt variations of a design by refining and sorting issues (Lawson 1997). Those meanings that are attached to shapes in a Sketch are ambiguous and dynamic, allowing opportunities for change as a result of new interpretations (Purcell and Gero 1998). However, unlike Idea Sketches, *Study Sketches* may often focus upon certain aspects of a design in more detail. In this sense, they provide opportunity to evolve the important aspects of a potential design, but still at a level of ambiguity that provides room for interpretation and exploration. Thus, *Study Sketches* are often visual design representations used for investigating the appearance and visual impact of ideas; such as aspects of geometric proportion, configuration, scale, layout and mechanism. A *Study Sketch* may express design ideas at a higher fidelity than the Idea Sketch, but still represent design intent at a low level of fidelity in order to stimulate exploration, albeit with a focus upon a particular aspect or design feature (Figures 4.7 and 4.8).

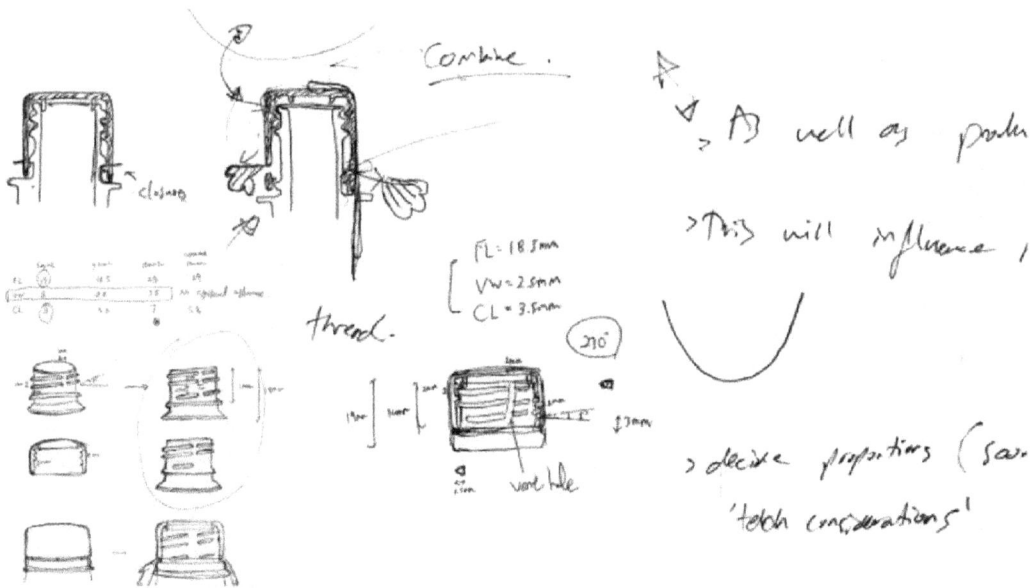

FIGURE 4.7
Examples of Study Sketches (Lee and Self, 2017).

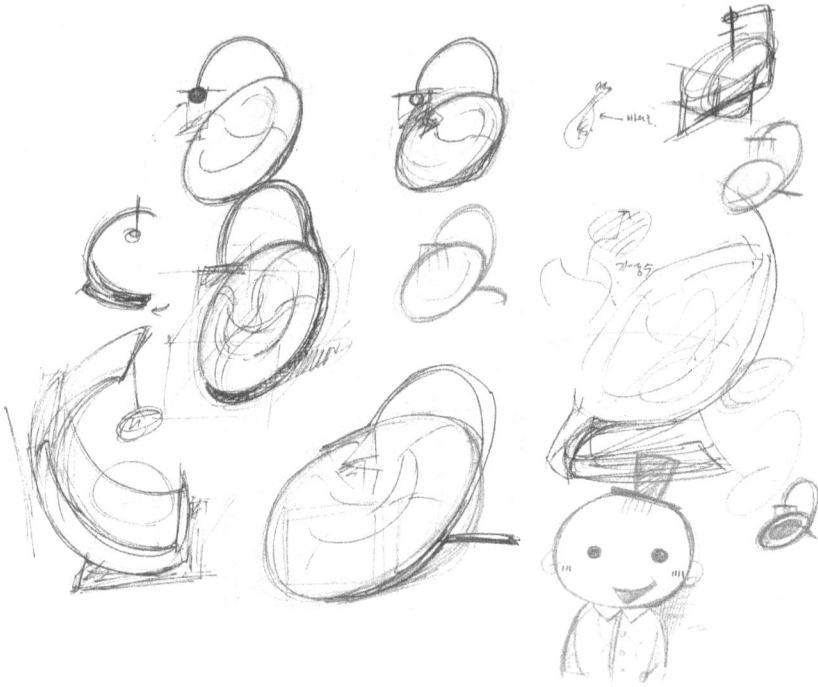

FIGURE 4.8
Examples of Study Sketches (Lee and Self, 2017).

4.4 Referential Sketch

According to Graves (1977), *Referential Sketches* or Storing Sketches (Ullman et al. 1990), are used to record observations and insights. Another use is to capture visual references that serve as inspiration (Olofsson and Sjölén 2005). *Referential Sketches* are 2D visual design representations used as a diary to record observations for future reference or as a metaphor. In this sense, they often act as a record to recall details and for inspiration about potential design directions going forward. They are a visual reference to aid memory and may stimulate ideas towards possible solutions as designers reflect upon their relevance for a current project, emergent design idea or as stimulation for design ideation and development (Figures 4.9 and 4.10).

Referential Sketches act to scaffold for design thinking as stimulation for cognition towards potential solution ideas, their various features and requirements. As *Referential Sketches* offer visual cues for reference and inspiration, they are a type of representation that does not aim to depict or express the design itself, but rather for considering broader issues related to context, design vision, identify or point towards important issues to consider. *Referential Sketches* therefore may not be assessed in terms of fidelity (their distance from a depiction/description of a final design solution). They may, in fact, provide images and references that only obliquely relate to any potential design solution at various levels of design and finality.

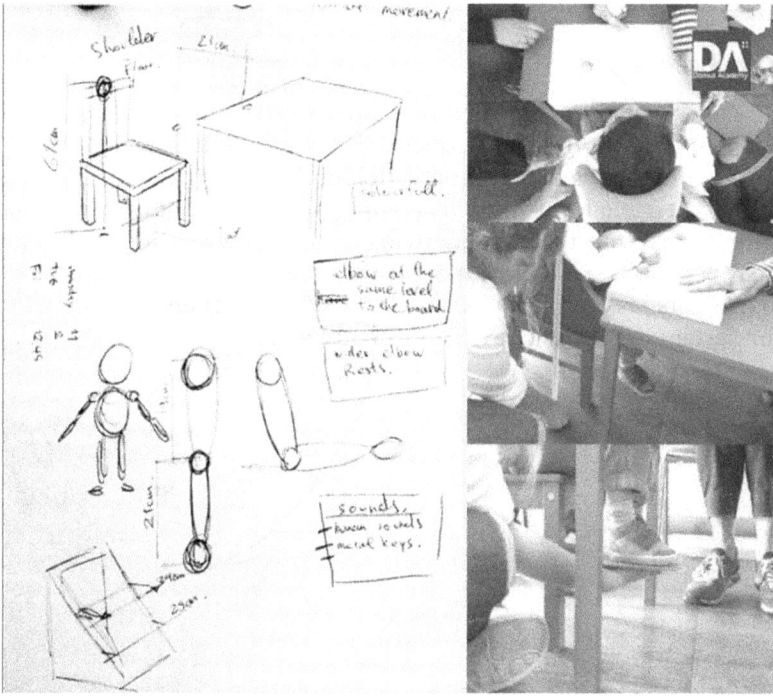

FIGURE 4.9
Referential Sketches (Aligizakis, n.d.).

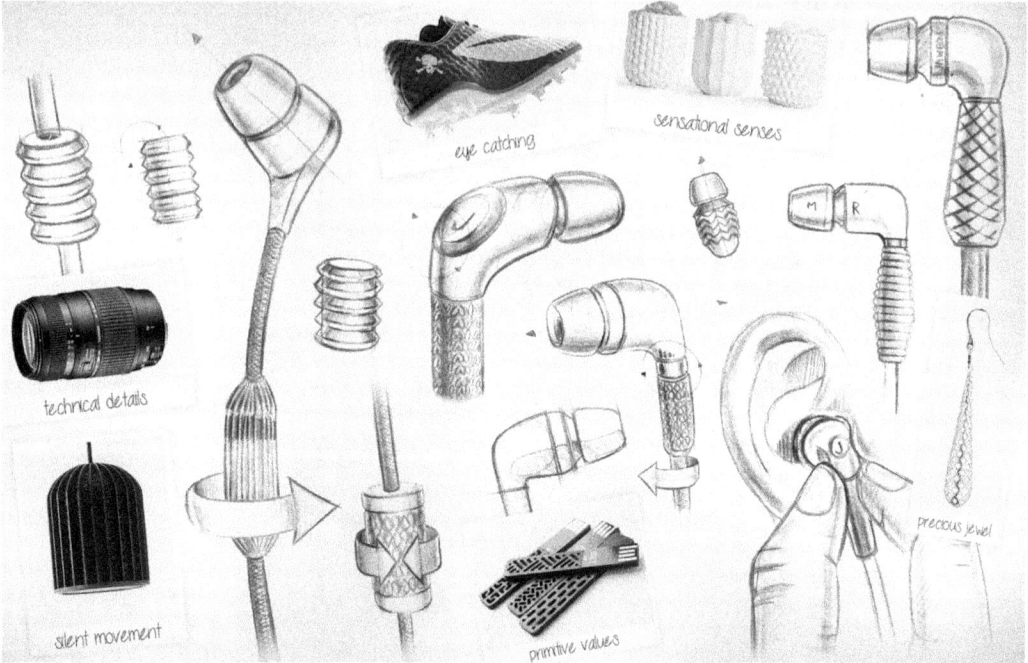

FIGURE 4.10
Images of Reference Sketches (Shah, n.d.).

4.5 Memory Sketch

These private sketches keep a record of thoughts and steps that have been taken, serving as an extension of memory (Do et al. 2000). While other sketches are used to develop concepts, *Memory Sketches* capture thoughts to retain information and to make such information easily accessible for further development (Van der Lugt 2005). *Memory Sketches* serve to distribute thinking through archiving of thoughts and ideas. These representations may then be consulted as prompts for memory and recall, supporting discussion between stakeholders, as well as personal records for designers themselves to track key features or important details of the design (Figure 4.11).

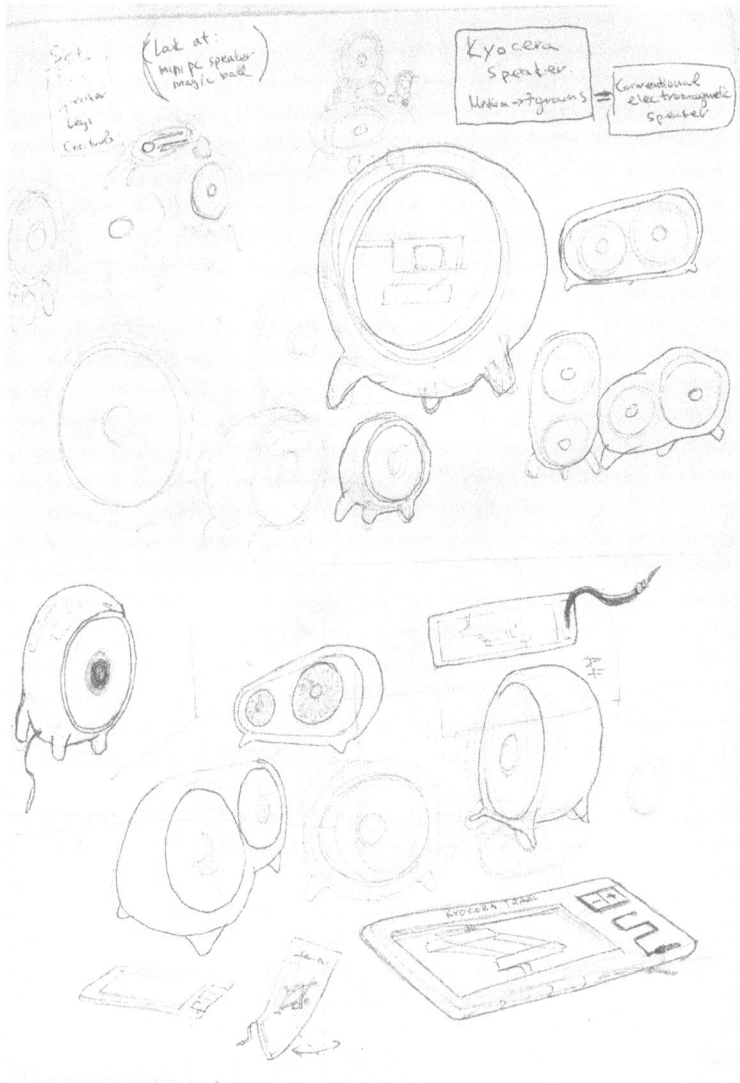

FIGURE 4.11
Memory Sketches (Aligizakis, n.d.).

Memory Sketches are 2D visual design representations that help users recall thoughts and elements from previous work, which often include notes and text annotations. As sketches, they are often produced at a lower level of fidelity, dependent upon what information is stored and for whom. They may also be more or less ambiguous in their expression of design intent, with ambiguity dependent on the stage they are used (i.e. *Concept, Concept Development* and *Detail Design*).

4.6 Coded Sketch

Coded Sketches employ the use of symbols to illustrate a principle (Tjalve et al. 1979b). Although they are similar to diagrams, *Coded Sketches* are icon-based, hand-drawn and use only a limited set of symbols. They are often used to establish relations between elements within a product or product-service systems. *Coded Sketches* are 2D visual design representations that categorise information to show an underlying principle or a scheme. *Coded Sketches* aim to approximate interaction or connections between elements in a more symbolic manner. They do not represent the form or mechanical function of a design, but rather express relations between elements or as part of a wider system of elements (Figure 4.12). As *Coded Sketches* do not express the physical or functional properties of a design (i.e. form and user interactions), they cannot be described in terms of fidelity derived from a distance from a final design solution. *Coded Sketches* are often used in practice to establish relationships between elements of a design and are symbolic in nature, rather than iconic in their expression of design intent; as is more usual in other types of sketch representation.

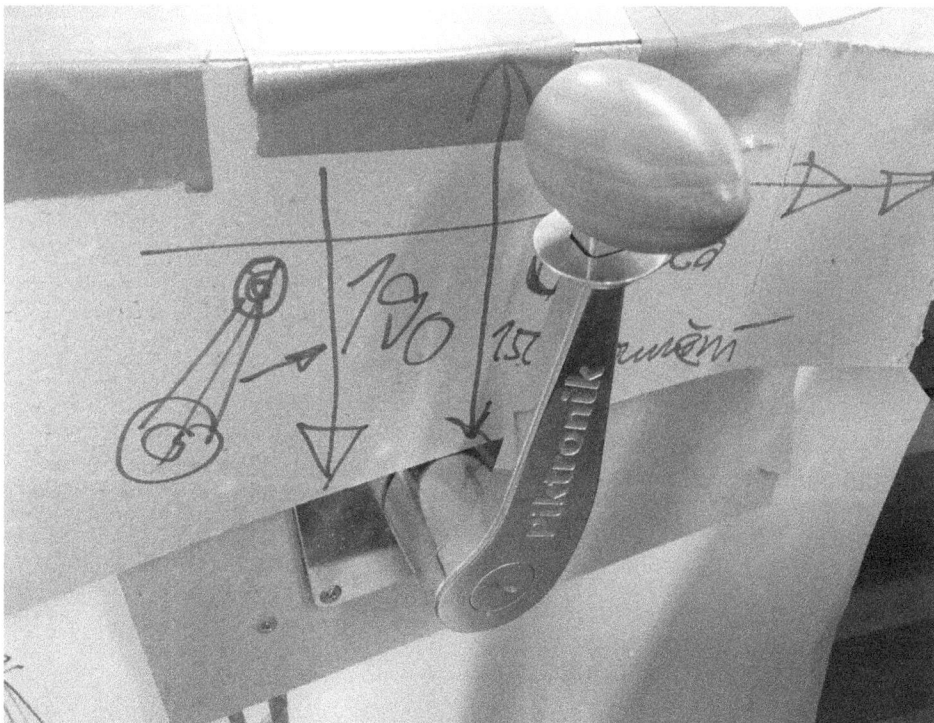

FIGURE 4.12
A Coded Sketch used on a existing mechanism (Prototypum, n.d.).

4.7 Information Sketch

These sketches are widely used by industrial designers to explain the form, function and structure of a concept to stakeholders and clients for evaluation (Van der Lugt 2005). They encourage discussion and a common understanding of the design idea among the team (Ferguson 1992). *Information Sketches*, then, aim to express and communicate design intent to others. Unlike more conceptual sketch types, they are often produced to explicate design detail that may be largely defined or developed prior to representation through an *Information Sketch*. They provide opportunities to communicate information related to design detail to other stakeholders. Colour and text annotations allow details to be explained clearly, and at a higher level of fidelity as well as adding realism to convey the design intent across the group. They are also known as Explanatory Sketches (Eissen and Steur 2008), Communication Sketches (Raudebaugh and Newcomer 1999), Pitching Sketches (Pavel 2005) or Talking Sketches (Ferguson 1992).

Information Sketches are 2D visual design representations that allow stakeholders to understand the designer's intentions by explaining information clearly, providing a common graphical setting (Figures 4.13 and 4.14). Due to these requirements, *Information Sketches* are often less ambiguous in their expression of intent, with a reduced opportunity for interpretation.

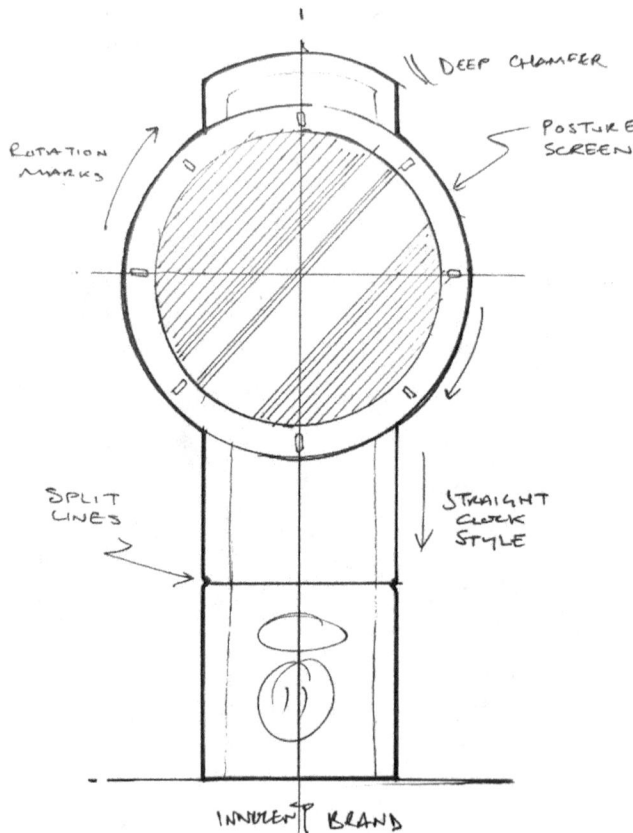

FIGURE 4.13
An Information Sketch exploring design detail (Choudhury, n.d.).

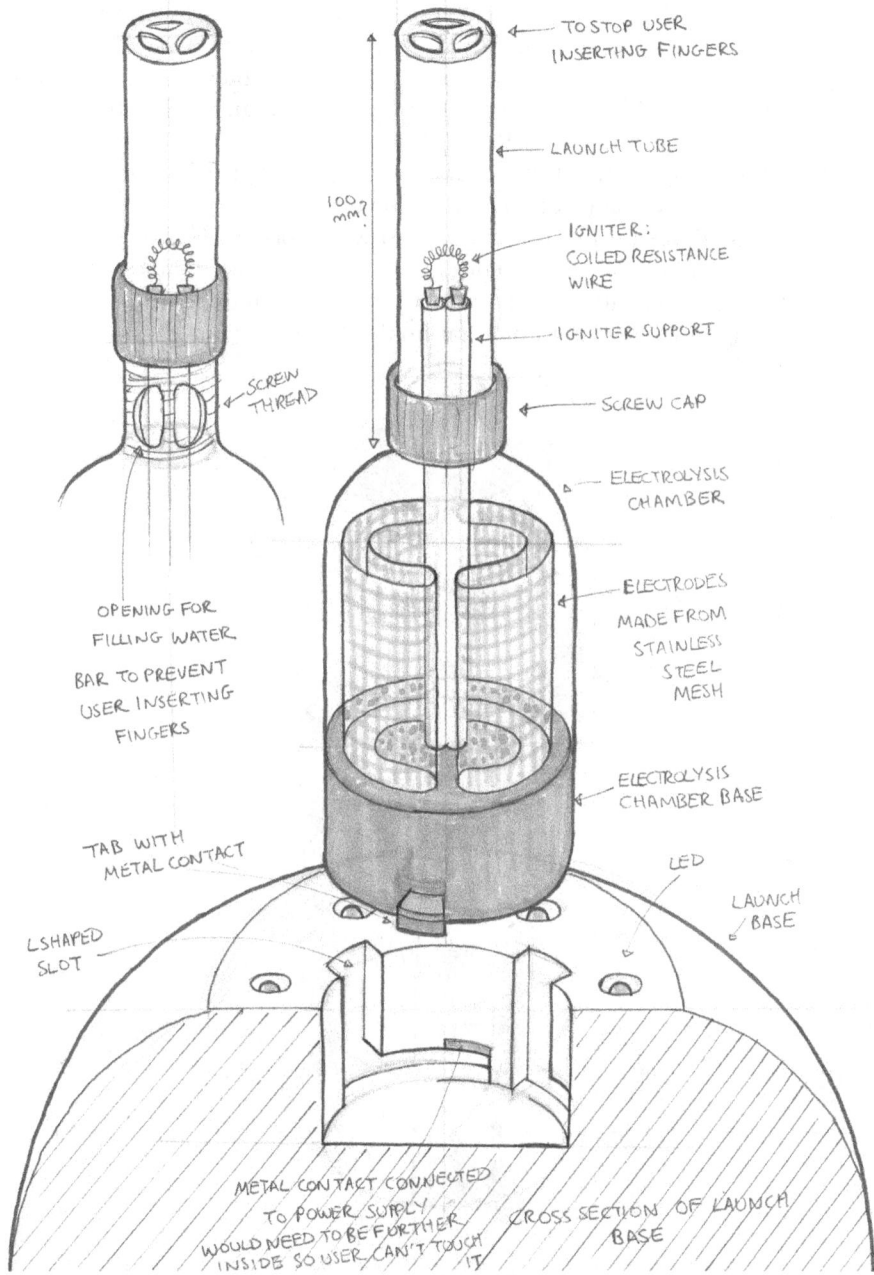

FIGURE 4.14
Information Sketches exploring further design detail (Churn, n.d.).

4.8 Renderings

Renderings involve the application of colour and tone to express a design's form and CMF (*Colour Material Finish*) as realistically as possible. The high level of realism reduces ambiguity and enables the viewer to better understand key features of the design (Evans 2002). *Renderings* are usually produced in perspective views and created either with manual media such as markers, or digitally (Goldschmidt 1992; Garner 2006). In terms of expression of form and CMF, they are often produced to higher levels of fidelity in the communication of intent. While *Renderings* may provide some room for interpretation, through the ambiguity of form and CMF, they are primarily used to persuasively express the aesthetic beauty of a potential design. *Renderings* are also known as Sketch Renderings (Evans 2002) or First Concepts (Monahan and Powell 1987).

Renderings are visual design representations showing formal proposals of design concepts that involve the application of colour, tone and detail for realism. They are often increasingly realistic, representing intent at a fidelity aimed at expressing a more realistic representation of intent, closer to a potential final design outcome (Figure 4.15).

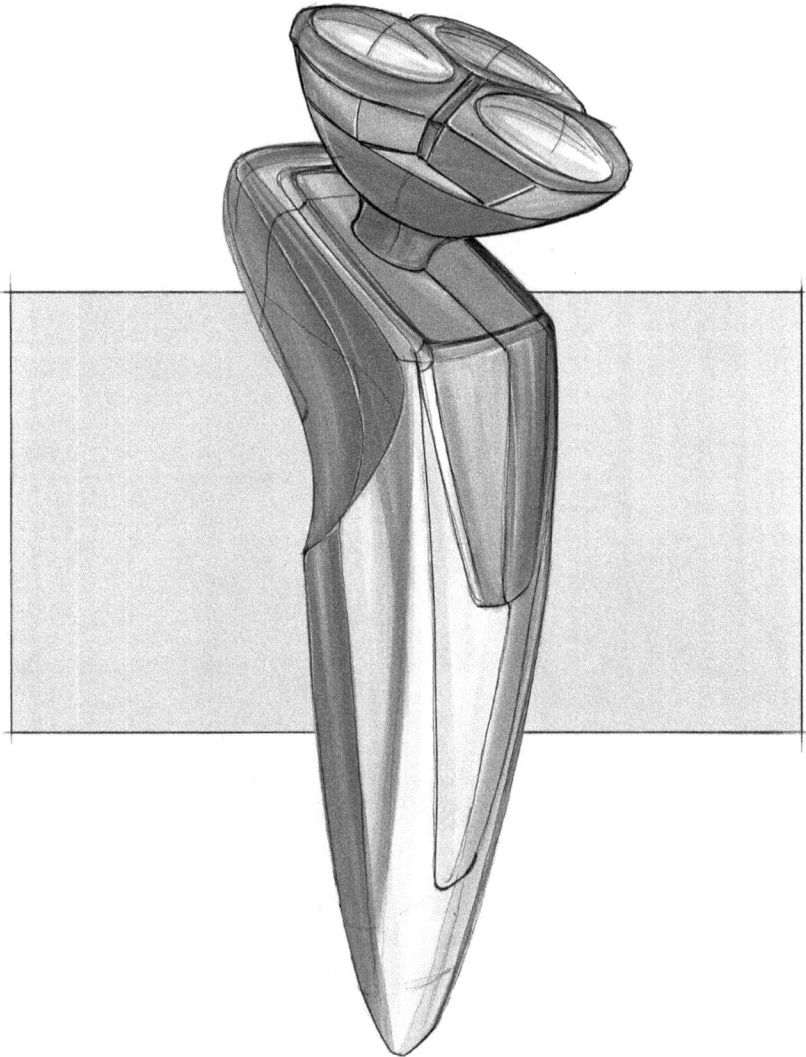

FIGURE 4.15
A Rendering used to explore the form of a product (Anderson, 2018).

4.9 Inspiration Sketch

These are highly form-orientated visuals that illustrate a design concept in detail. The purpose of an *Inspiration Sketch* is to influence an audience and to sell the idea by using artistic qualities to convey an emotion or a theme. In this sense, they represent design intentions at a high level of fidelity in terms of an overall aesthetic direction. Although in reality, much design detail may remain unresolved. Although they may be time-consuming to produce, they express qualities that are hard to achieve with 3D CAD modelling (Olofsson and Sjölén 2005). As the main aim is to convey the feel of a product, *Inspiration Sketches* may not be accurate in their fidelity and detail of design expression. *Inspiration Sketches* are also known as Visionary Drawings (Lawson 1997) or Emotional Sketches (Ling 2006a). They are form-orientated, visual design representations used to communicate the look or feel of a product by setting the tone of a design, brand or product range (Figures 4.16–4.19). They provide clarity in terms of aesthetic direction but may be more ambiguous in the expression of design detail.

FIGURE 4.16
Inspiration Sketches (Prototypum, n.d.).

FIGURE 4.17
Another example of an Inspiration Sketch (Prototypum, n.d.).

FIGURE 4.18
An Inspiration Sketch (Walters, n.d.).

FIGURE 4.19
Another example of an Inspiration Sketch to indicate the surface profile (Walters, n.d.).

4.10 Prescriptive Sketch

According to Pipes (2007), *Prescriptive Sketches* are created during the development stages of the design process prior to a more detailed General Arrangement Drawing. They show key dimensions in a freehand orthographic projection with three views drawn to scale (Bertoline 2002). They are used for checking details in preparation for the physical or CAD Model and are also known as Specification Sketches (Pavel 2005). In their representation of intent, they attempt to communicate design detail unambiguously, in line with a requirement for clear communication to help finalize design intent. Although *Prescriptive Sketches* are informal visual design representations, they aim to communicate design decisions and general technical information such as dimensions, material and finish to a higher level of detail and fidelity (Figures 4.20 and 4.21).

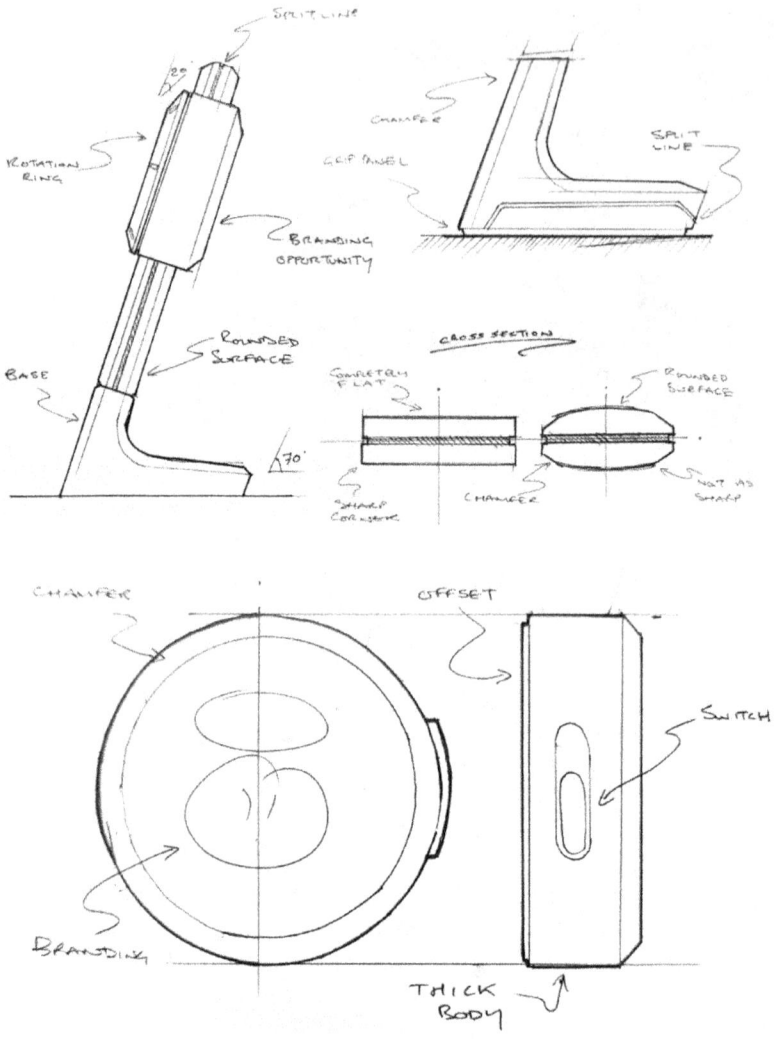

FIGURE 4.20
A Prescriptive Sketch used to communicate specific design detail (Choudhury, n.d.).

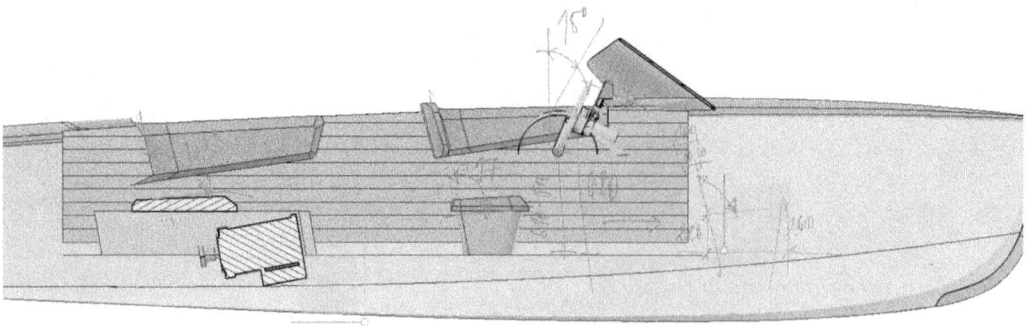

FIGURE 4.21
Another example of a Prescriptive Sketch (Prototypum, n.d.).

4.11 Summary

Chapter 4 has explored the role and use of sketch representations as ubiquitously used in product design and development. The chapter commenced with a review of sketching as a representational tool in design, its historic roots and relation to design thinking, communication in design and the design process (Section 4.1). The use of design sketching was traced back as a requirement to express potential solutions prior to manufacturing as a result of a separation of craft during the industrial revolution. Its role as support for the conceptualisation of design intent and its further development were also considered. Ambiguity in the representation of design intent was also discussed in terms of its implication for communication of ideas between stakeholders.

Various approaches to the definition and taxonomy of sketching types were discussed. These included the classification of sketches in terms properties of the representation (i.e. *line-thickness, use of colour, composition*), the purposes of a Ssketch (*to explore, to persuade, to define*), or the stage in the process a sketch type may be applied (i.e. *Concept Design, Concept Development, Detail Design*). The chapter then described and discussed different types of sketch (Sections 4.2–4.10) as a holistic list of sketch representations used in practice: Section 4.2, Section 4.3, Section 4.7, Section 4.8, Section 4.9 and Section 4.10.

Sketch types were discussed in relation to the degree of ambiguity, from the highly ambiguous *Idea Sketch*, to the specifics of the *Prescriptive Sketch*. Likewise, sketch representations were considered in terms of fidelity between sketch representation and the final design outcome; from a wider gap between fidelity of expression as sketch representation (i.e. *Study Sketch*), to representation that aims to communicate the detail of a design (*Prescriptive Sketch*). Discussion also mapped the type of sketch representation to the phase of the design process most often used (*Concept Design, Concept Development, Detail Design*).

Sections 4.4–4.6 referred to three types of sketch representation that defied classification through either a phase-in process or the related character of the representation assessed in terms of ambiguity and fidelity. Instead, these sketch types are used throughout a process of design to act as a scaffold for memory (*Referential Sketch, Memory Sketch*), or to provide an opportunity to explore relationships between elements (*Coded Sketch*) but in an abstract manner, often supported by symbolic visualizations of relations.

In providing a description of various sketch types and their relation to process and representational needs across conceptual, developmental and detail design activities, we have attempted to establish a relationship between the types of sketch used and the evolving requirements of design practice. As discussed in Chapter 2 (*Design Thinking through Representation*), the types of sketches employed appear to depend on and are related to problem-solving and creative cognition engaged at various phases of a design process. Conceptual design requiring the expression of design intent that may provide an opportunity for greater interpretation of meaning (i.e. *Study Sketch*). In contrast, definitive sketch representations can support a working thought of more localized issues and requirements within the design (*Prescriptive Sketch*). These, in particular, may be used as means to communicate intent to other stakeholders.

References

Baskinger, M. (2008) Pencils Before Pixels: A Primer in Hand-Generated Sketching. *Interactions* March–April, Vol. 15(2), pp. 28–36.

Bertoline, G. R. (2002) *Introduction to Graphics Communications for Engineers* (2nd ed.). New York: McGraw Hill.

Buxton, B. (2007) *Sketching User Experiences - Getting the Design Right and the Right Design*. San Francisco: Morgan Kaufmann.

Chen, H.-H., You, M., et al. (2003) The Sketch in Industrial Design Process. *Proceedings of the 6th Asian Design Conference*: Integration.

Do, E. Y.-L. (2005) Design Sketches and Sketch Design Tools. *Knowledge-Based Systems*, Vol. 18, pp. 383–405.

Do, E. Y.-L., Gross, M. D., et al. (2000) Intentions in and Relations Among Design Drawings. *Design Studies*, Vol. 21(5), pp. 483–503.

Dictionary of Art Terms (2003) *Thames and Hudson World of Art*. In: Lucie-Smith, E. (ed.). London: Thames and Hudson.

Ehrlenspiel, K. and Dylla, N. (1993) Experimental Investigation of Designers Thinking Methods and Design Procedures. *Journal of Engineering Design*, Vol. 4(3), pp. 201–212.

Eissen, K. and Steur, R. (2008) *Sketching: Drawing Techniques for Product Designers*. Singapore: Bis Publishers/Page One Publishing.

Evans, M. A. (2002) *The Integration of Rapid Prototyping within Industrial Design Practice* (Staff Thesis). Loughborough: Department of Design and Technology, Loughborough University.

Ferguson, E. S. (1992) *Engineering and the Mind's Eye*. Cambridge, Massachusetts: MIT Press.

Fish, J. C. (1996) *How Sketches Work-A Cognitive Theory for Improved System Design*. PhD dissertation. Loughborough University of Technology. Quoted in: Do, E., Gross, M., D., Neiman, B. and Zimring, C. (2000) Intentions in and Relations Among Design Drawings. *Design Studies*, Vol. 21(5).

Garner, S. (2006) *Modelling Workbook 1: T211 Design and Designing Workbook 1 Technology Level 2* (2nd ed.). Milton Keynes: The Open University Press.

Graves, M. (1977) The Necessity for Drawing: Tangible Speculation. *Architectural Design*, Vol. 6, pp. 384–394. Quoted in: Do, E. (2002) Drawing Marks, Acts, and Reacts: Toward a Computational Sketching Interface for Architectural Design. *Artificial Intelligence for Engineering Design, Analysis and Manufacturing*, (16), pp. 149–171.

Goldschmidt, G. (1992) Serial sketching: Visual problem solving in designing. *Cybernetics and Systems*, Vol. 23, pp. 191–219. Quoted In: Cardella, M., Atman E., Cynthia J. and Adams, R., S. (2006) Mapping Between Design Activities and External Representations for Engineering Student Designers. *Design Studies*, Vol. 27(1).

Kojima, T., Matsuda, S., et al. (1991) Models and Prototypes. *Tokyo: Graphic-Sha Publishing*. In: Evans, M. (2002) *The Integration of Rapid Prototyping within Industrial Design Practice* (Staff Thesis). Loughborough: Department of Design and Technology Loughborough University.

Lawson, B. (1997) *How Designers Think – the Design Process Demystified*. Oxford: Architectural Press. Quoted in: Persson, S. and Anders, W. (2003) Relational Modes between Industrial Design and Engineering Design – a Conceptual Model for Interdisciplinary Design Work. *Proceedings of the 6th Asian Design International Conference*, Tsukuba.

Ling, B. (2006a) *Design Sojourn – Do I have to be Able to Draw Well to be a Good Designer?* http://www.designsojourn.com/index.php/2006/12/06/do-i-have-to-be-able-todraw- well-to-be-a-gooddesigner/. Accessed on 17 June 2007.

Ling, B. (2006b) *Tips on How to Improve your Drawing Ability*. http://www.designsojourn.com/2006/12/08/tips-on-how-to-improve-your-drawing-ability/. Accessed on 14 December 2006.

McGown, A., Green, G., et al. (1998) Visible Ideas: Information Patterns of Conceptual Sketch Activity. *Design Studies*, Vol. 19(4), pp. 431–453.

Monahan, P. and Powell, D. (1987) *Advanced Marker Techniques*. London: MacDonald and Co. In: Evans, M. (2002) *The Integration of Rapid Prototyping within Industrial Design Practice* (Staff Thesis). Loughborough: Department of Design and Technology Loughborough University.

Moyer, D. (2007) *Napkin Sketches 101*. http://www.steelcase.com/uk/files/3d025ccaa4924ac6b67e 7d20a89b35b3/Full%20versio n%20of%20this%20story.pdf. Accessed on 3 February 2007.

Olofsson, E. and Sjölén, K. (2005) *Design Sketching*. Sundsvall, Sweden: Keeos Design Books AB.

Pavel, N. (2005) *The Industrial Designer's Guide to Sketching*. Trondheim: Tapir Academic Press.

Pipes, A. (2007) *Drawing for Designers*. London: Laurence King Publishing.

Prats, M., Lim, S., et al. (2009) Transforming Shape in Design: Observations from Studies of Sketching. *Design Studies*, 30(5), pp. 503–520.

Purcell, A. T. and Gero, J. S. (1998) Drawings and the Design Process. *Design Studies*, Vol. 19(4), pp. 389–430.

Raudebaugh, R. A. and Newcomer, J. (1999) *Visualization, Sketching and Freehand Drawing for Engineering Design*. Mission, Kansas: Schroff Development Corporation.

Rodriguez, W. (1992) *The Modelling of Design Ideas – Graphics and Visualization Techniques for Engineers*. Singapore: McGraw-Hill Book Company.

Schön, D. A. (1983) *The Reflective Practitioner: How professionals think in action*. London: Temple Smith.

Stacey, M. and Eckert, C. (2003) Against Ambiguity. *Computer Supported Cooperative Work*, Vol. 12, pp. 153–183.

Tjalve, E., Andreasen, M. M. et al. (1979b) *Engineering Graphic Modelling – A Workbook for Design Engineers*. London: Butterworth and Co.

Tovey, M., Porter, S., et al. (2003) Sketching, Concept Development and Automotive Design. *Design Studies*, Vol. 24(2), pp. 135–153.

Tovey, M. (1989) Drawing and CAD in Industrial Design. *Design Studies*, Vol. 10(1), pp. 24–39.

Ullman, D. G., Wood, S., et al. (1990) The Importance of Drawing in the Mechanical Design Process. *Computer and Graphics*, Vol. 14(2), pp. 263–274.

Van der Lugt, R. (2005) How Sketching can Affect the Idea Generation Process in Design Group Meetings. *Design Studies*, Vol. 26(2), pp. 101–122

Verstijnen, I. M., Hennessey, J. M., et al. (1998) Sketching and creative discovery. *Design Studies*, Vol. 19(4), pp. 519–546.

Yang, M. C. and Cham, J. G. (2007) An Analysis of Sketching Skill and Its Role in Early Stage Engineering Design. *Transactions of the ASME*, Vol. 129, pp. 476–482.

5

Drawings

5.1 Drawing as design representation

A drawing is a more formal arrangement of lines that determines a particular form (Dictionary of Art Terms 2003). When compared with Sketches, Drawings are highly structured to formalise and verify aspects of the design (Herbert 1993; Robbins 1994; Goel 1995). Ullman et al. (1990) also clarified that drawings are made in accordance with a set of rules and are drafted with mechanical instruments or a CAD system to scale; whereas Sketches are done almost free-hand and are often not to scale. Due to a greater adherence to formal rules and conventions, drawings are often less ambiguous in their expression of design intent. They are also often produced at a decreased distance from the expected design solution, or feature a design solution at increased fidelity. For example, an exploded drawing to describe the General Arrangement of parts (Figure 5.1). In classifying drawings, Fraser and Henmi (1994) analysed architectural drawings and grouped them as *Referential Drawings*, *Diagrams*, *Design Drawings*, *Presentation Drawings* and *Visionary Drawings*. As with sketch representations, these classifications were derived from the purpose of the drawing. However, unlike Sketches, Drawings as design representations appear more objectively classifiable due to their greater adherence to rules and conventions.

A formal definition of drawn representation was proposed by Tjalve et al. (1979) who defined drawings as the modelled properties of a design such as the structure, form, material, dimension, surface, etc. Drawings were also coded with use of formal symbols such as coordinates, graphical symbols, types of projection, etc. (ibid). Discussing the purpose of various drawings, Tjalve et al. (ibid), describe them as serving as a record to analyse and check details, as well as a more formal communication medium between the designer and the manufacturer (Ullman et al. 1990). Besides the type of projection, drawings include the use of colour and dimensions to provide specific information (Yang 2003; Song and Agogino 2004), such as the drawing produced by Rebecca Churn (Figure 5.2).

In addition, there exists more formal conventions to prescribe the features of different drawn representations, such as BS 8888; an established standard developed by the British Standards Institution (BSI) for technical product documentation, geometric product specification, geometric tolerance specification and engineering drawings. Similarly, ASME Y14.5 is a standard published by the American Society of Mechanical Engineers (ASME) that establishes rules, symbols, definitions, requirements, defaults and recommended practices for stating and interpreting geometric dimensions and tolerances for engineers, manufacturers and designers. In particular, technical and part drawings may attempt to specify design in a way that reduces or eliminates interpretation. In terms of the design process, drawings are more often used during *Concept Development* and *Detail Design*, as the design process pivots from reflective exploration through to a clearer articulation of the features and details of the

DOI: 10.1201/9781003227694-5

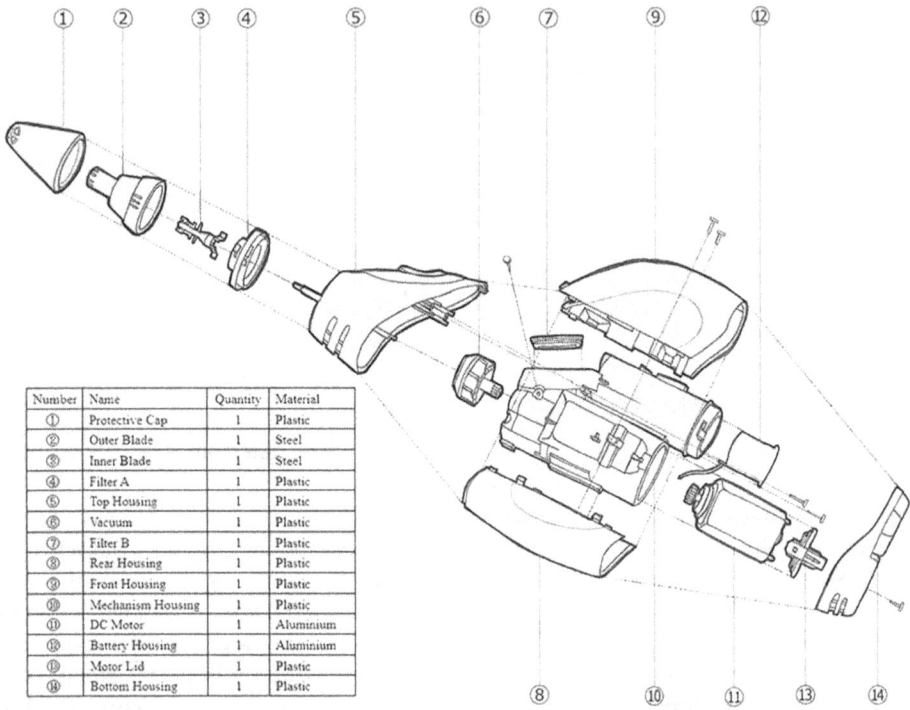

Number	Name	Quantity	Material
①	Protective Cap	1	Plastic
②	Outer Blade	1	Steel
③	Inner Blade	1	Steel
④	Filter A	1	Plastic
⑤	Top Housing	1	Plastic
⑥	Vacuum	1	Plastic
⑦	Filter B	1	Plastic
⑧	Rear Housing	1	Plastic
⑨	Front Housing	1	Plastic
⑩	Mechanism Housing	1	Plastic
⑪	DC Motor	1	Aluminium
⑫	Battery Housing	1	Aluminium
⑬	Motor Lid	1	Plastic
⑭	Bottom Housing	1	Plastic

FIGURE 5.1
General Arrangement Drawing showing parts in an exploded view (Hayashi, n.d.).

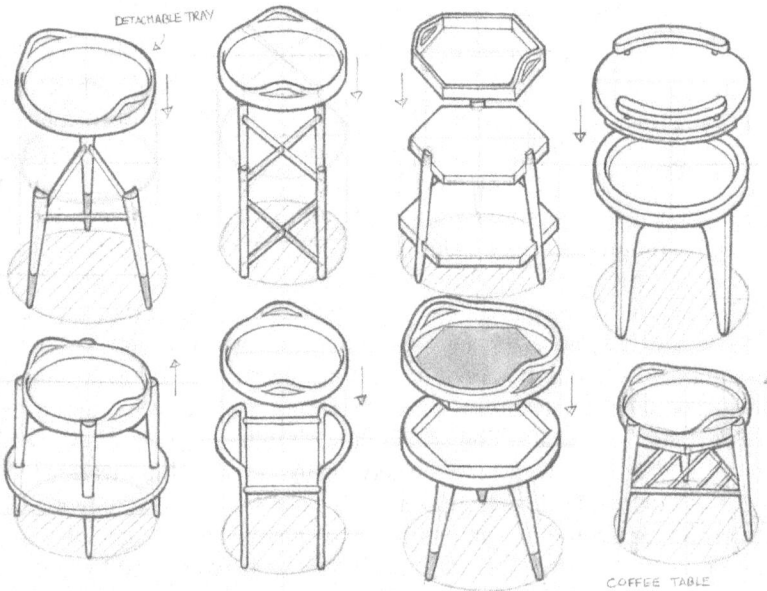

FIGURE 5.2
Drawings used to indicate function of parts (Churn, n.d.).

proposed solution. Both in terms of their adherence to more formalised conventions, and relation to design development and detailing, drawn representations depart from the sketch types discussed in Chapter 4. Although, like sketch representations, drawings are produced through various media such as digital drawing tablets, paper and pencil, design drawings may also be differentiated in terms of their purpose of use. Sketch representations are often produced to support exploration of possibilities, providing greater opportunity for interpretation through ambiguity of expression. In contrast, drawings more often aim to communicate the detail of design intent across various stakeholders as the design process moves further towards development, detailing and implementation.

The following sections describe nine types of drawn representation often used in practice. Classification broadly follows the use in the design process, from *Concept Design* (Section 5.2) to *Detail Design* (Section 5.10).

5.2 Concept Drawing

Also known as Layout Drawings (DTI 1992), *Concept Drawings* are used by industrial designers to define the form of a design and to show how the finished product would appear in an orthographic view. Usually several of these drawings are used in internal discussions to evaluate possible proposals (Tovey 1989). *Concept Drawings* are 2D visual design representations that show the design proposal in colour with orthographic views and precise lines (Figure 5.3). Figure 5.4 shows a Concept Drawing produced by the Samsung Design Europe Studio to represent colour, material and finishing, to be implemented on the reference model. Like sketches, they offer some room for reflective interpretation. However, they also express design intent at a higher level of fidelity compared to an Information Sketch for example (see Section 4.7).

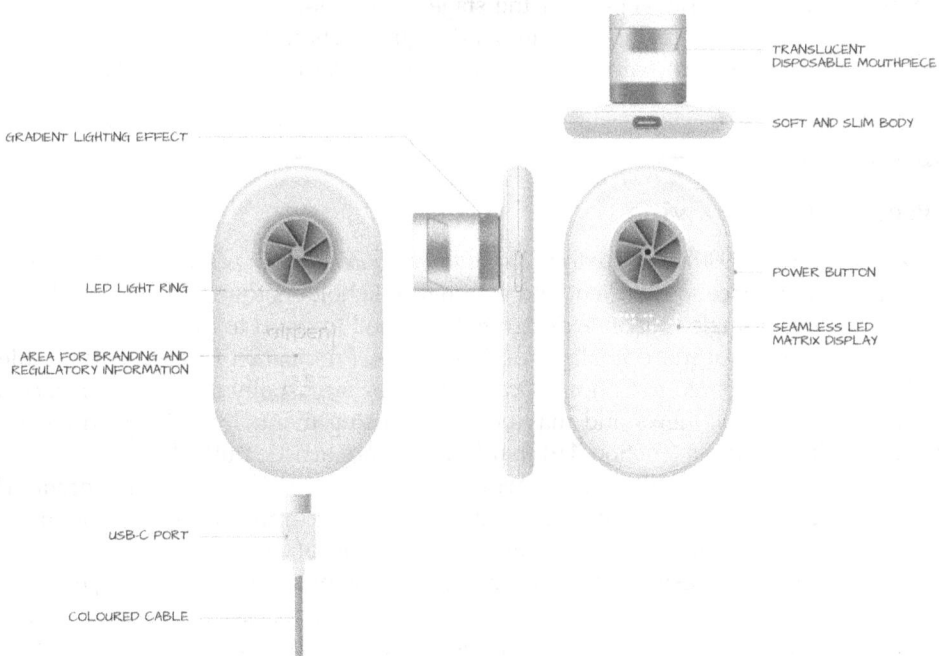

FIGURE 5.3
Concept Drawings (Lad, n.d.).

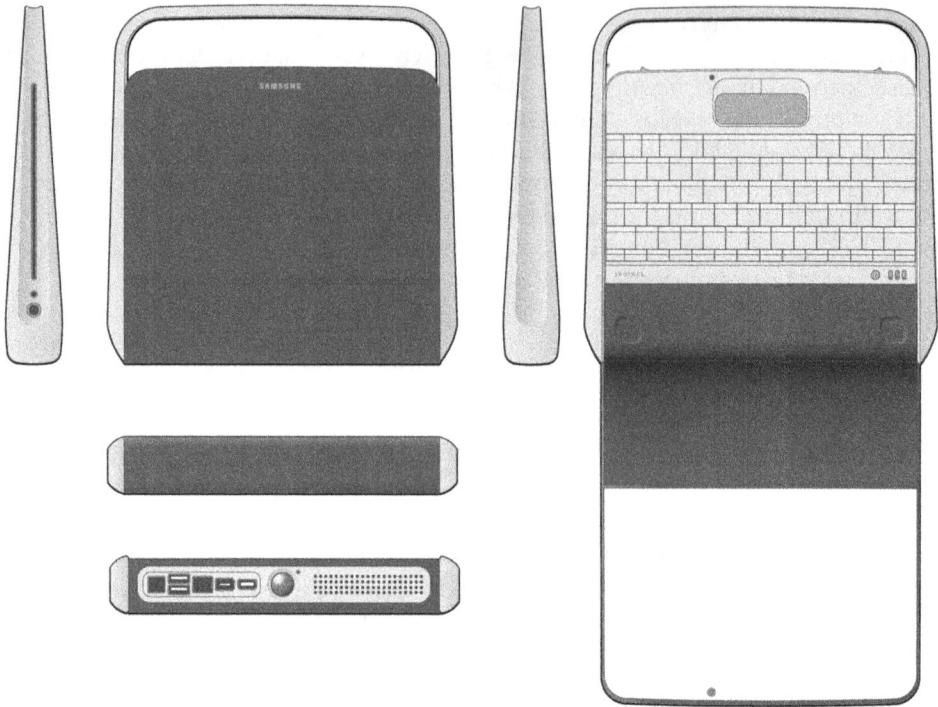

FIGURE 5.4
Another example of a Concept Drawing (Hill, n.d.; Samsung Design Europe Studio).

In terms of the stage in the design process, *Concept Drawings* may be used during concept design as an outcome or deliverable of the stage in process. These drawings may also be shared to support decision making towards a design direction, framing or reframing of the design problem and solution pairing or the expression of a more holistic aesthetic direction.

5.3 Presentation Drawing

According to Powell (1990) and Buxton (2007), *Presentation Drawings* are used to sell an idea and to inspire confidence in the client and external stakeholders towards concept ideas. The outcome is usually a single workable design to be carried forward to the next phase (*Concept Development*) to work out more finalised design details. *Presentation Drawings* offer a higher level of realism as compared to Concept Drawings. They are usually drawn in perspective as opposed to orthographic views and may be created using manual media or on computer; unlike Inspiration Sketches (Section 4.9) that have a more artistic outlook.

Presentation Drawings are also more formal in their expression of design intent. They represent design ideas with decreasing ambiguity of communication. Here, purpose moves away from interpretation to a fuller presentation of design intent. Although the detail of a propositional design solution has yet to be established, the *Presentation Drawing* aims to provide a vision of the final product (Figure 5.5). This then acts as means to establish and confirm a design direction for the overall aesthetic content, form, use, function and/or CMF (Colour Material Finish design).

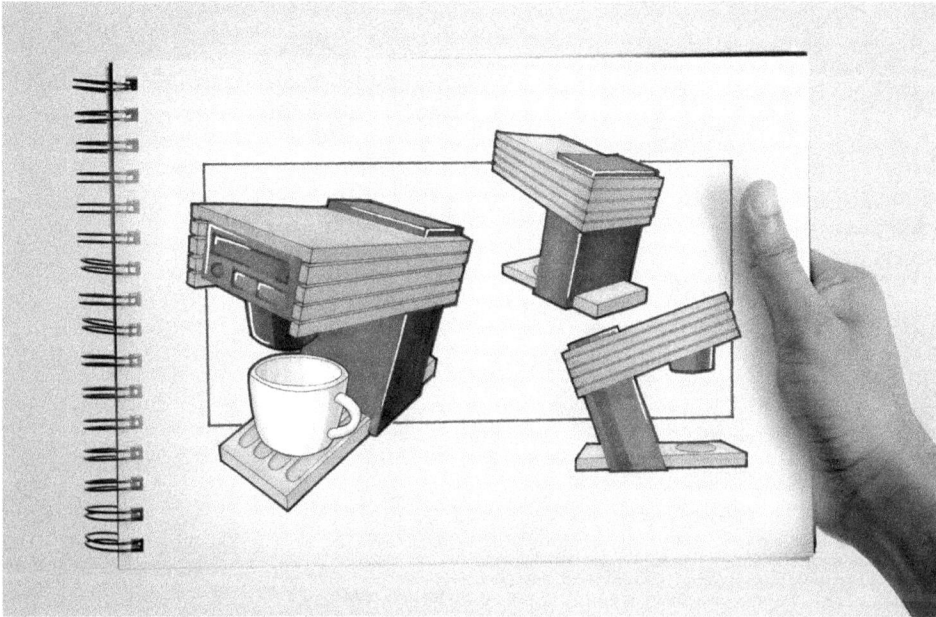

FIGURE 5.5
Presentation Drawings in sketchbook (Choudhury, n.d.).

5.4 Scenarios and storyboards

These 2D visual design representations aim to explain a concept by illustrating the possible settings of a product, user and/or an environment and context of use. They are often employed along with text to enhance understanding. (Olofsson and Sjölén 2005). They may take the form of a timeline to describe stages of a product's use (Pavel 2005). *Scenarios and Storyboards* may also be further defined as 2D visual design representations to suggest user and product interaction, and to portray its use in the context of artefacts, people and relationships.

Design representations as *Scenarios and Storyboards* approximate use and function as part of a wider product context. In this expression, they aim to clearly articulate a proposed solution related to the user experience. *Scenarios and Storyboards*, as focused upon use and context, may be employed during *Concept Design* and *Concept Development* stages of a design process. As their aim is often to provide a more holistic overview of use and context, they may communicate ideas with some ambiguity in the expression of product detail and/or contextual elements. However, as a communicative tool, *Scenario and Storyboards* also aim to clearly communicate how the product experience locates within a broader context of use (Figures 5.6 and 5.7). In this sense, the drawings are more symbolic in their communication of intent as they provide indicators towards activities, emotions and experiences derived from product engagement and within a particular scenario of use.

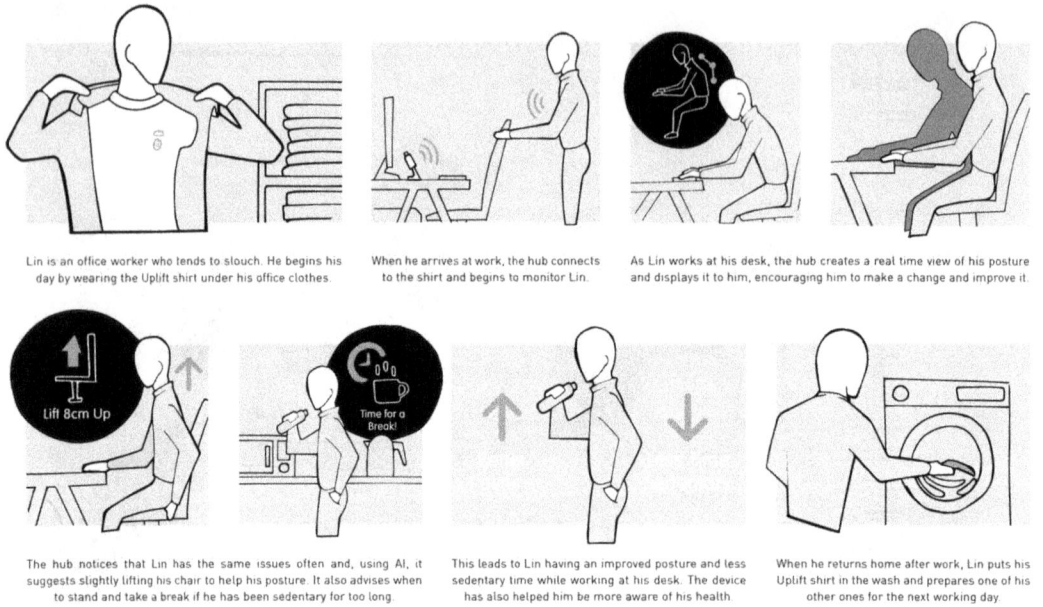

Lin is an office worker who tends to slouch. He begins his day by wearing the Uplift shirt under his office clothes.

When he arrives at work, the hub connects to the shirt and begins to monitor Lin.

As Lin works at his desk, the hub creates a real time view of his posture and displays it to him, encouraging him to make a change and improve it.

The hub notices that Lin has the same issues often and, using AI, it suggests slightly lifting his chair to help his posture. It also advises when to stand and take a break if he has been sedentary for too long.

This leads to Lin having an improved posture and less sedentary time while working at his desk. The device has also helped him be more aware of his health.

When he returns home after work, Lin puts his Uplift shirt in the wash and prepares one of his other ones for the next working day.

FIGURE 5.6
Scenarios and Storyboards to explain the use of a product (Choudhury, n.d.).

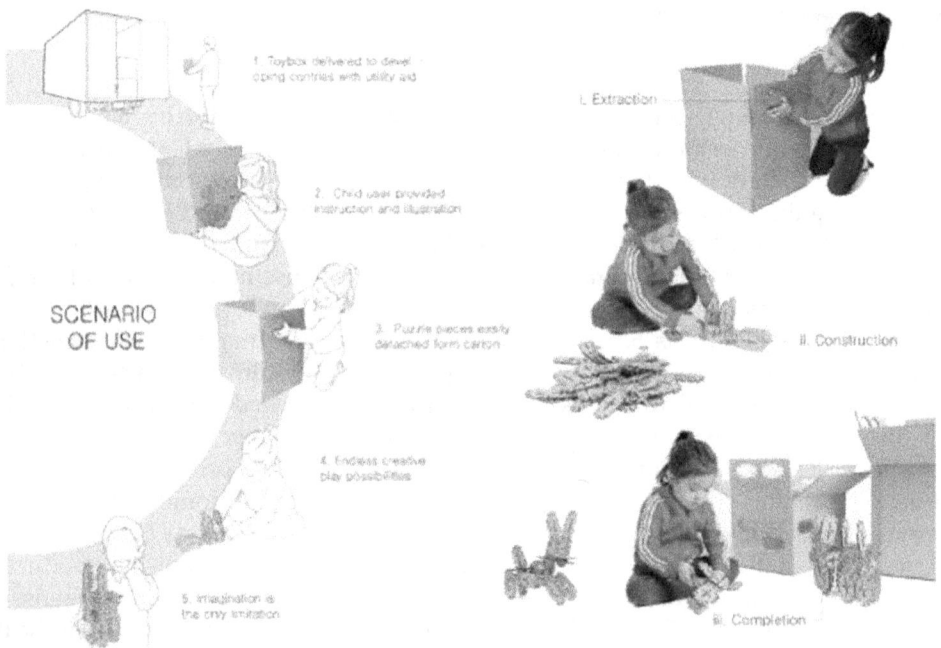

FIGURE 5.7
Another example of Scenarios and Storyboards (Self, 2018).

5.5 Diagrammatic Drawing

The purpose of a *Diagrammatic Drawing* is to group data visually so that the information can be clearly understood (Blackwell 1997). They are also used to show the structure and relationships of components in a system. Most *Diagrammatic Drawings* are represented with simple geometric elements such as arrows, solid lines and hatching lines to illustrate the principle or operation of the system (Do et al. 2000). For clarity, aesthetic form is often omitted (Lawson 1997). *Diagrammatic Drawings* are also known as *Diagrams* or *Schematic Drawings* (Tovey 1989), with more common *Diagrammatic Drawings* including mechanical, hydraulic, pneumatic, electronic and electrical symbols or elements to record functional structures of the product (Tjalve et al. 1979). Larkin and Simon (1987) noted that because the information within *Diagrammatic Drawings* is indexed, they may only be useful to those who understand the codes, and so diagrams follow a formal structure that must be learnt. To those unfamiliar with these codes, *Diagrammatic Drawings* are ambiguous and obscure. However, for those who understand the codes and conventions used, diagrams often

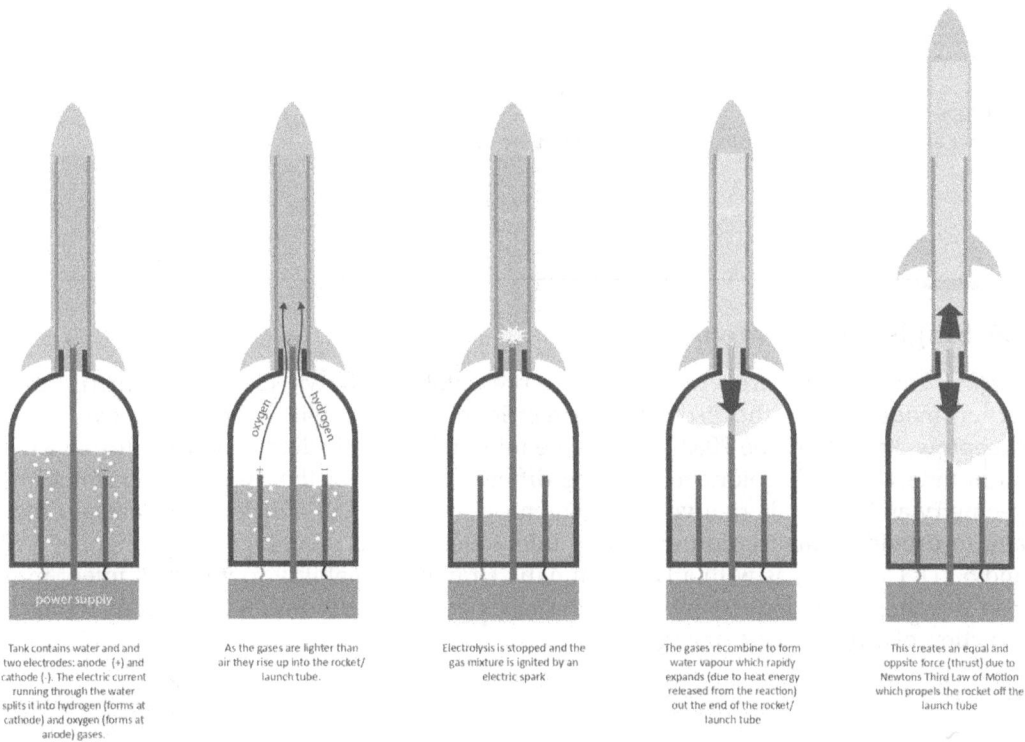

Tank contains water and and two electrodes: anode (+) and cathode (-). The electric current running through the water splits it into hydrogen (forms at cathode) and oxygen (forms at anode) gases.

As the gases are lighter than air they rise up into the rocket/ launch tube.

Electrolysis is stopped and the gas mixture is ignited by an electric spark

The gases recombine to form water vapour which rapidly expands (due to heat energy released from the reaction) out the end of the rocket/ launch tube

This creates an equal and oppsite force (thrust) due to Newtons Third Law of Motion which propels the rocket off the launch tube

FIGURE 5.8
Diagrammatic Drawings indicating technical function (Churn, n.d.).

aim to represent design intention towards specific design detail unambiguously, thereby aiming to avoid interpretation.

Diagrammatic Drawings are often produced as 2D visual design representations that show the underlying principle of a solution idea or represent relationships between objects, represented with simple geometric elements (Figures 5.8 and 5.9).

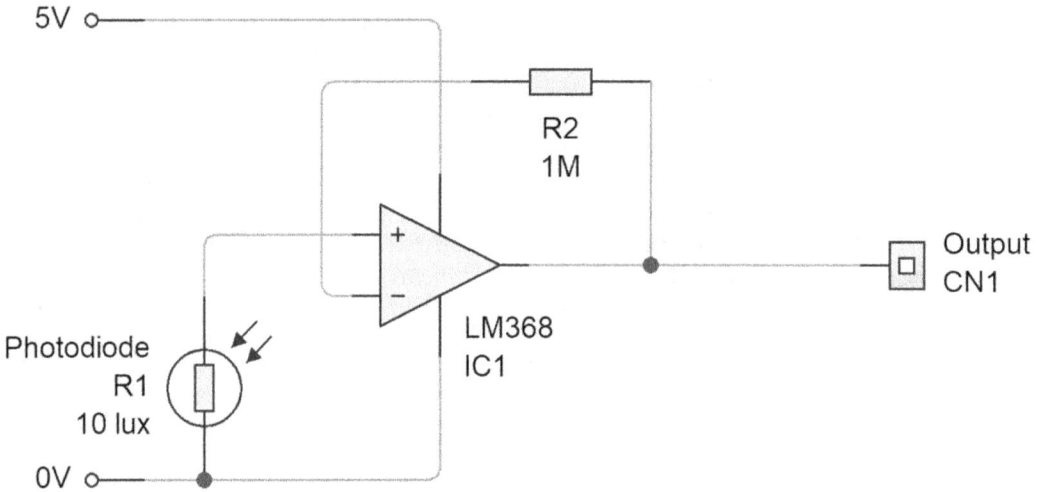

FIGURE 5.9
Another example of a Diagrammatic Drawing (Mortimer, n.d.).

5.6 Single-View Drawing

Single-View Drawings are visual design representations drawn in an axonometric projection made up of either isometric, trimetric, diametric, oblique or perspective views (Lueptow 2000; Bertoline 2002). They have minimal aesthetic detail and are illustrated as an outline with little colour to describe different aspects of the design, to examine the geometry and show alternative arrangements (Do et al. 2000). *Single-View Drawings* thus aim to provide a description of design intent clearly and unambiguously (Figures 5.10 and 5.11). However, as with a Diagrammatic Drawing, an ability to interpret intent may also depend upon understanding of the rules and conventions that underlie the construction of *Single-View Drawings*.

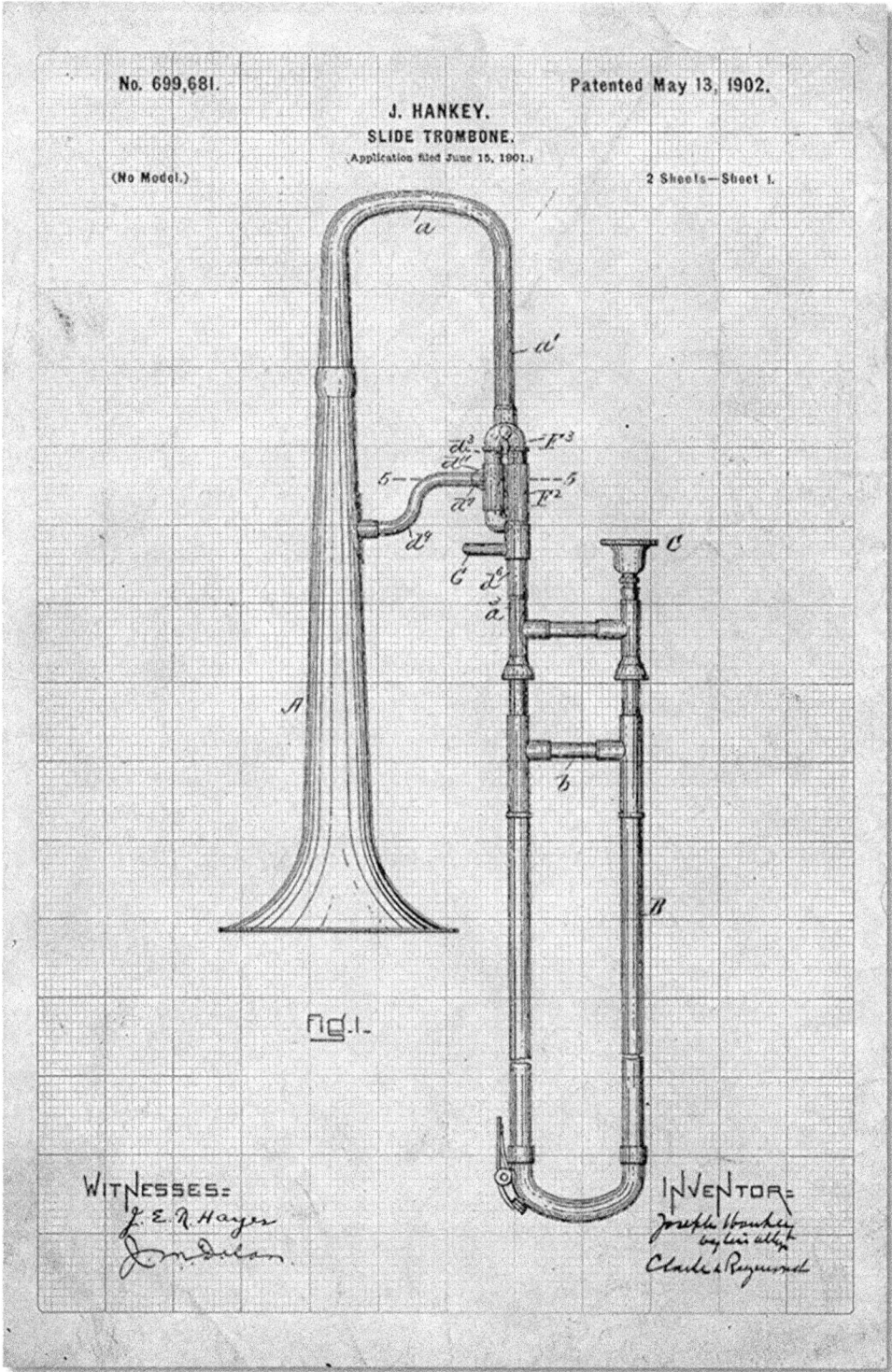

FIGURE 5.10
A Single-View Drawing (Aligizakis, n.d.).

before stage   after stage

Min.
33.2°C

Heat = 0,4 W
Normal convection
Material = Al 6061
if Ta = 30°C

36.2°C
Max.

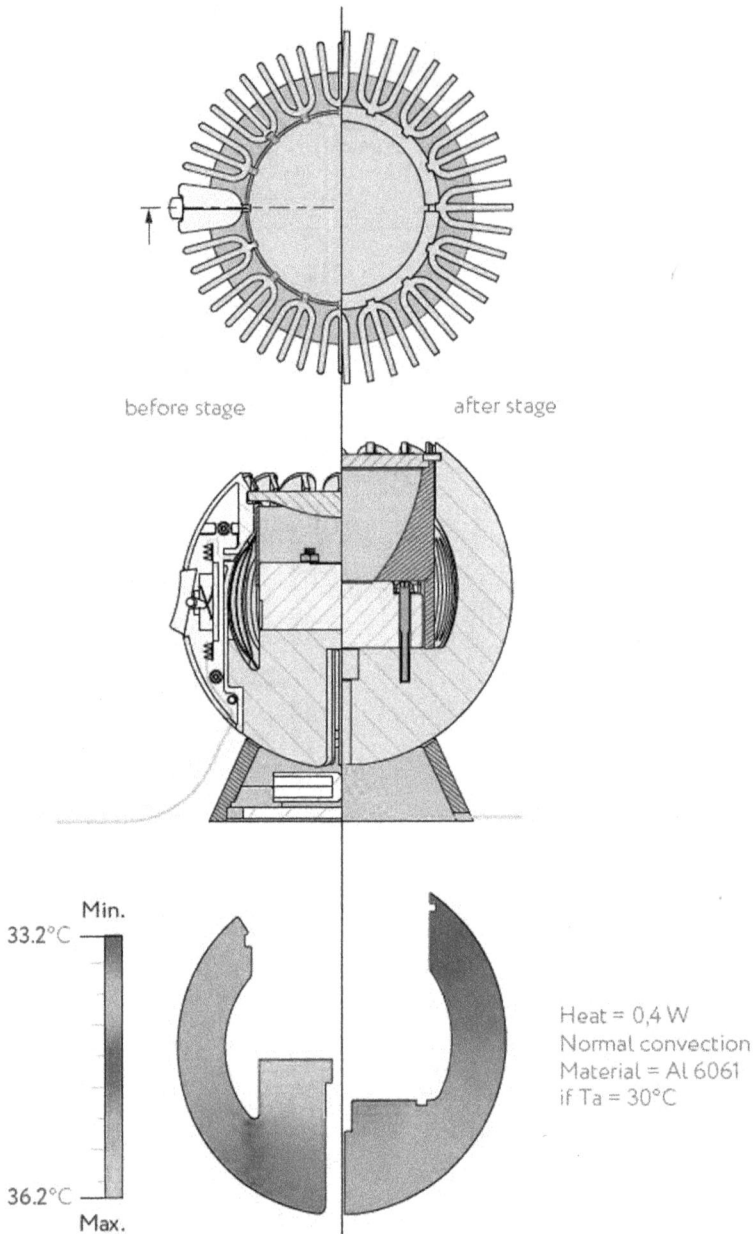

FIGURE 5.11
Another example of a Single-View Drawing (Dermachi, n.d.).

5.7 Multi-View Drawing

Multi-View Drawings consist of projections to describe a product in 2D (Pavel 2005). Also known as Orthographic Projections, they are a formal system used to describe the features and geometry of a product through three coordinated orthogonal planes made up of plan view, front elevation and end elevation (Raudebaugh and Newcomer 1999; Bertoline 2002). There are two types of projections for *Multi-View Drawings*. A first-angle projection consists of a plan view and the front face drawn immediately above it and the end elevation to the right. In a third-angle projection, one elevation is placed below the plan, with the end elevation to the left of the first elevation. *Multi-View Drawings* are 2D visual design representations employed through first or third angle projections (Figure 5.12). In all cases, as with Diagrammatic Drawings, these drawings often abide by formal conventions in order to clearly and precisely articulate the part geometry and design detail. *Multi-View Drawings* exist to provide more detailed description of a solution in relation to part geometry and exact dimension and specification. As such they are often used at a Detail Design phase to communicate intent towards a final product outcome. However, *Multi-View Drawings* omit information towards other aspects (i.e. colour, material, aesthetic).

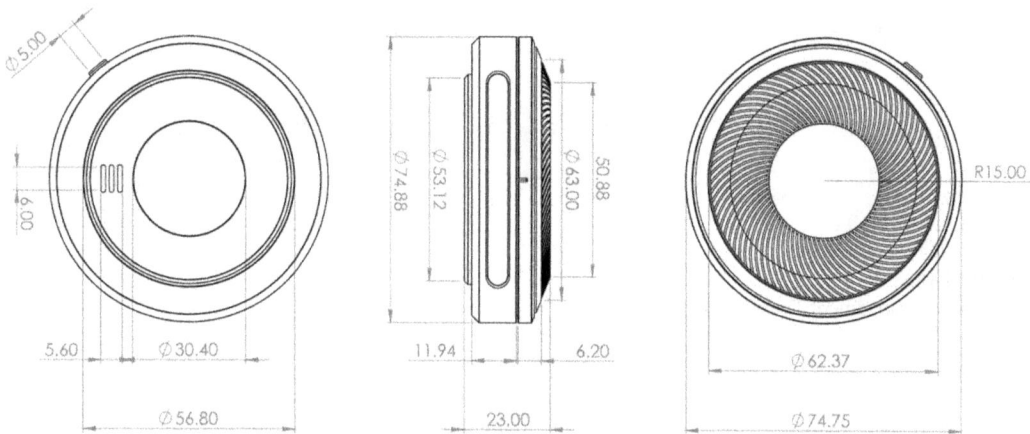

FIGURE 5.12
A Multi-View Drawing (Choudhury, n.d.).

5.8 General Arrangement Drawing

Once a concept has been approved, the next step is to produce a *General Arrangement Drawing* (also known as a GA Drawing), or Model Making Drawings (DTI 1992). At the concept development stage, the design has a refined layout with fixed dimensions. They are created prior to a Technical Drawing (see below) and represent an overview of the design and how the parts are put together (Powell 1990). As compared with Prescriptive Sketches, *GA Drawings* are more formal in their incorporation of a multi-view drawing, dimensions, parts list, sub-assemblies, drawing angles and break lines (Martin 1989; DTI 1992). When colour and shading are applied, *General Arrangement Drawings* become a powerful communication tool that can be used for discussions with non-technical members (Powell 1990). *GA Drawings* are often used by model makers for creating Appearance Models.

General Arrangement Drawings are 2D visual design representations that embody the refined design but omit the internal details (Figures 5.13 and 5.14). As such they represent the form and composition of a design to a high degree of fidelity but can omit other aspects of design (i.e. internal components).

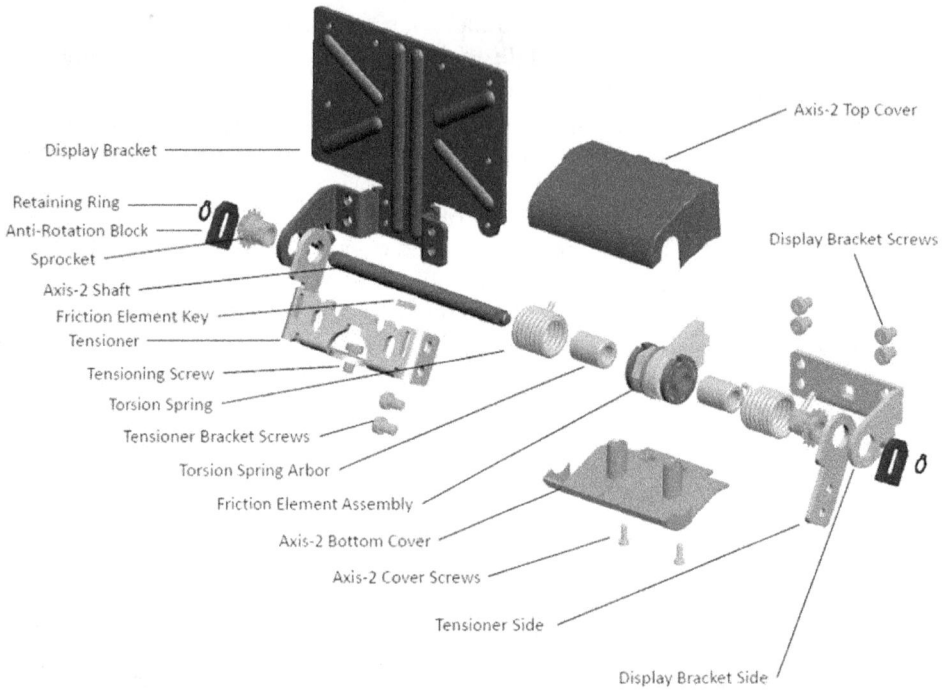

FIGURE 5.13
A General Arrangement Drawing (Chan, n.d.).

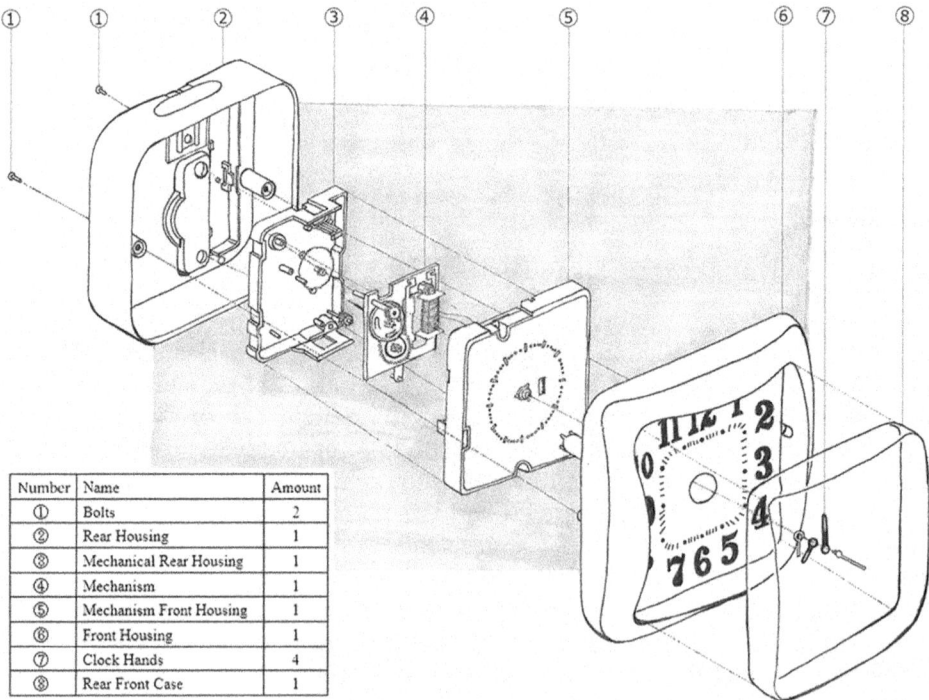

Number	Name	Amount
①	Bolts	2
②	Rear Housing	1
③	Mechanical Rear Housing	1
④	Mechanism	1
⑤	Mechanism Front Housing	1
⑥	Front Housing	1
⑦	Clock Hands	4
⑧	Rear Front Case	1

FIGURE 5.14
Another example of a General Arrangement Drawing (Hayashi, n.d.).

5.9 Technical Drawing

Technical Drawings are formal 2D visual design representations used to define, specify and graphically represent the built object and to cover every detail for manufacture. They represent the last stage of the design development process where the design is ready for manufacture. *Technical Drawings* therefore attempt to represent design unambiguously, and with consideration for and communication of holistic design detail. However, as with other design drawings used to represent intent at a *Detail Design* phase, an ability to interpret intent can depend upon a formal understanding of the rules and conventions governing their use.

Technical Drawings may be created by manual drafting or through a computer. Also known as Documentation Drawings (Raudebaugh and Newcomer 1999) or Production/Working Drawings (Bertoline 2002). They are formalised, complete and standardised, showing the material specification, parts list, manufacture, finish and assembly details (Figures 5.15 and 5.16). Design representation as a *Technical Drawing* may also be used for organising and calculating the production costs involved in the manufacture of a product (Tjalve et al. 1979). To ensure clarity and consistency, most *Technical Drawings* conform to industry standards such as the British Standards Institution (BSI) BS8888 standard with guidelines that define, specify and graphically represent products. In the USA, the American equivalent is the ASME Y14.5M standard for dimensioning and tolerancing

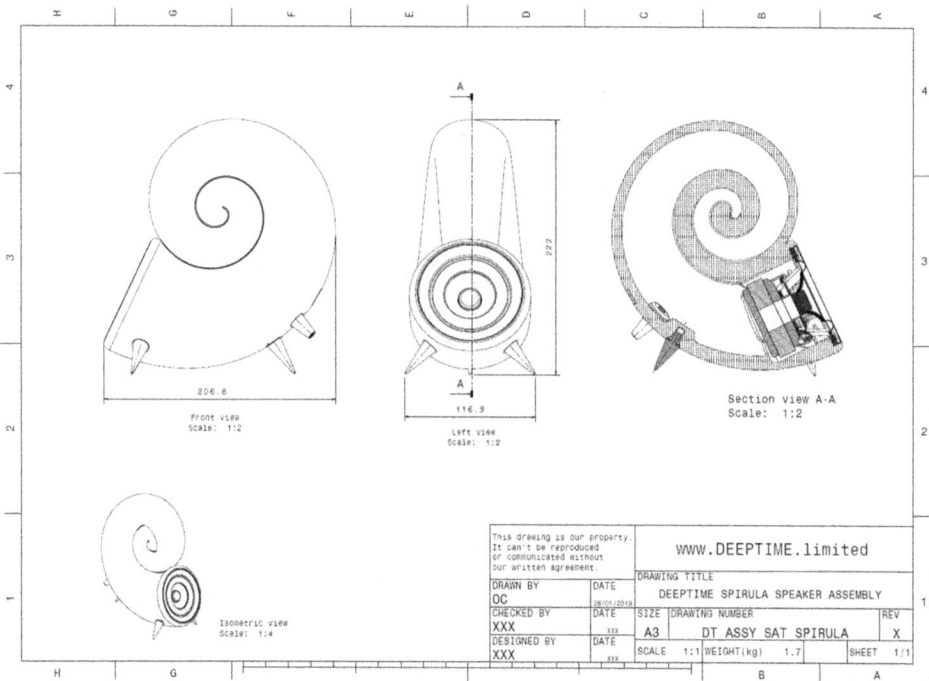

FIGURE 5.15
Technical Drawing providing a representation of a product (Deeptime, 2019).

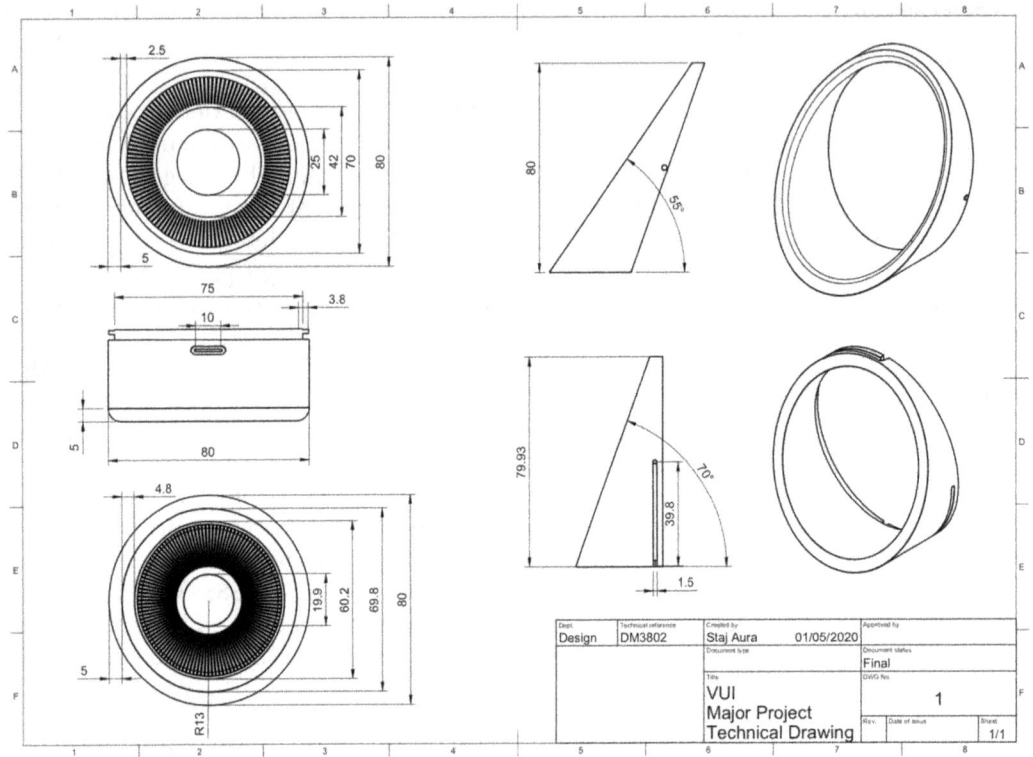

FIGURE 5.16
Another example of a Technical Drawing (Aura, n.d.).

(Pipes 2007). Their clarity of communication is premised upon an ability to comprehend the rules and conventions that apply to their construction. *Technical Drawings* may also be used as a precursor to the act of prototyping. The experiments by Yang (2003) suggested that producing increased amounts of dimensioned drawings early in the design cycle could lead to a more positive design outcome.

5.10 Technical Illustration

These are representations created at the very end of the development process (*Detail Design*). Because orthographic projections or technical drawings may be too complex for a layman to understand, *Technical Illustrations* simplify the engineering details and highlight key features without omitting important information (Pipes 2007). To enhace communication of design detail, *Technical Illustrations* are accompanied with sections, cut-aways, ghosting and exploded views. Cut-aways show the inside of a product that may be, for example, hidden by a casing (Eissen and Steur 2008). Ghosting is another technique that makes an area transparent to show the internal components and keeping the overall form recognisable.

Although similar to Presentation Drawings, *Technical Illustrations* are used to explain the engineering aspects rather than to communicate design aesthetics (Bertoline 2002). They may be created with airbrush or on a computer and are used for instruction manuals, installation guides, maintenance manuals, catalogues, advertisements and in training books. As such, *Technical Illustrations* are 2D visual design representations that simplify the engineering details and highlight key features without omitting important information from the product. The artwork created by Sam Gwilt (Figure 5.17) showing the *Technical Illustration* of the camera was commissioned by Procreate, a digital software used for drawings, illustrations and painting. Due to their enhanced communication of design intent, *Technical Illustrations* often present a highly developed representation of a solution in detail. Their purpose is not to provide opportunity for interpretation of information through ambiguity of expression, but rather to communicate the specifics of refined design intent. In this sense, they are often produced at a high degree of fidelity intheir expression of form, material, colour and component architecture (Figure 5.18).

FIGURE 5.17
A Technical Illustration (Gwilt, n.d.).

FIGURE 5.18
Another example of a Technical Illustration (Walters, n.d.).

5.11 Summary

This chapter has discussed the role and use of drawing as representational media in design. The chapter commenced with a discussion of drawing as a means to express intent at various phases of the design process (Section 5.1). In this discussion, we provided a distinction between drawing and sketching related to drawing's purpose of use in practice. In particular, drawing as design representation, departing from many types of sketch representation, is employed to communicate intent to others. In this, drawings are described as being more detailed and less ambiguous in their expression of design ideas as compared to Sketches. This difference is particularly conspicuous in those types of Drawing that follow codes and conventions in their articulation of design detail (i.e. Sections 5.8 and 5.9).

Drawings as design representations were described as expressions of design intent often used in the communication of various design features. For example, the *Presentation Drawing* (Section 5.3) is employed to express and communicate an overall design aesthetic, including colour, material and finish. *Technical Illustrations* (Section 5.10) are used in the representation of engineering and manufacturing aspects such as part geometry and fit, and often prescribes to certain conventions of use.

Drawn design representations were also distinguished in their reduced use of ambiguity in the expression of design intent, compared to sketch representation. Related to their purpose of use as increasingly weighted towards communication, space for interpretation recedes and is replaced with a desire for the comprehensive communication of potential solution ideas. Likewise, fidelity of representation through drawing increases to

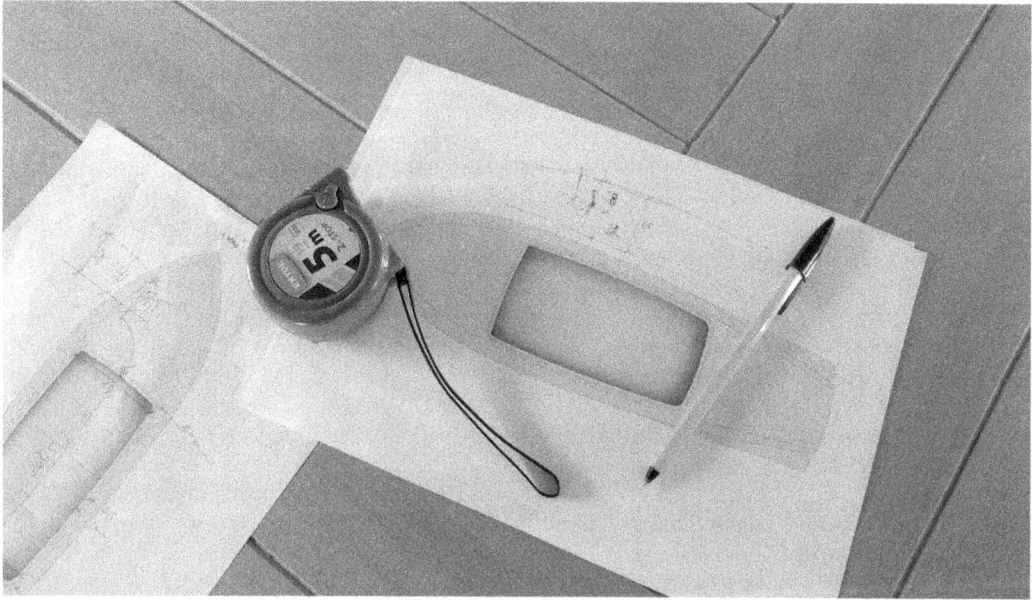

FIGURE 5.19
Drawings used in the design process (Prototypum, n.d.).

narrow the gap between drawn representation and a potential final design solution, or part thereof (Figures 5.19). Related to a reduction in ambiguity of expression and narrowing of representational fidelity, design drawings were discussed in terms of phase of the design process (i.e. *Concept Design, Concept Development, Detail Design*). Chapter 5 has identified and described nine types of Drawings (Sections 5.2–5.10). As with sketch representations (Chapter 4), our presentation of Drawings broadly align with their application at different phases of a design process: *Concept Design, Development Design* and *Detail Design*. However, as with sketch representations, some Drawings were identified as used across various phases (Sections 5.4 and 5.5).

Here we have aimed to provide a broad overview of the role and use of various Drawings as representations. We do not see our taxonomy of Drawing as exclusive. Instead, we provide a holistic account of various drawn representations, their relation to the design process and reasons for relationships between types of representation used and their ability to meet the varying demands of practice. Together with Chapter 4, the two chapters have documented 18 types of design representations used in practice. This then constitutes the most inclusive account of various drawn and sketched design representations to date.

In Chapters 6 and 7, we turn to design representations that approximate aspects of design intent in a physical form, and consider their role as means to express, explore and communicate design intent.

References

Bertoline, G. R. (2002) *Introduction to Graphics Communications for Engineers* (2nd ed.). New York: McGraw Hill.

Blackwell, A. F. (1997) Diagrams about Thoughts about Thoughts about Diagrams. In: Anderson, M. (1997) Reasoning with Diagrammatic Representations II, *AAAI 1997 Fall Symposium* held in Menlo Park, California pp. 77–84. Quoted in: Do, E. Y., Gross, M. D. and Zimring, C. (1999) Drawing and Design Intentions: An Investigation of Freehand Drawing Conventions in Design. *Proceedings Design Thinking Research Symposium* held in Cambridge, MA.

Buxton, B. (2007) *Sketching User Experiences - Getting the Design Right and the Right Design*. San Francisco: Morgan Kaufmann.

Dictionary of Art Terms (2003) *Thames and Hudson World of Art*. E. Lucie-Smith (Ed.). London: Thames and Hudson.

Do, E. Y.-L., Gross, M. D., et al. (2000) Intentions in and Relations among Design Drawings. *Design Studies*, Vol. 21(5), pp. 483–503.

DTI (1992) *Managing Product Design Projects*. London: Department for Trade and Industry.

Eissen, K. and Steur, R. (2008) *Sketching: Drawing Techniques for Product Designers*. Singapore: Bis Publishers/Page One Publishing.

Fraser, I. and Henmi, R. (1994) *Envisioning Architecture: An Analysis of Drawing*. New York: Van Nostrand Reinhold. Quoted in: Lawson, B. (1997) *How Designers Think - The Design Process Demystified*. Oxford: Architectural Press.

Goel, V. (1995) *Sketches of Thought*. Cambridge, MA: MIT Press.

Herbert, D. M. (1993) *Architectural Study Drawings*. New York: Van Nostrand Reinhold. Quoted in: Do, E. Y-L. (2002) Drawing Marks, Acts, and Reacts: Toward a Computational Sketching Interface for Architectural Design. *Artificial Intelligence for Engineering Design, Analysis and Manufacturing*, No. 16, pp. 149–171.

Larkin, J. H. and Simon, H. A. (1987) Why a Diagram Is (Sometimes) Worth Ten Thousand Words. *Cognitive Science Journal*, Vol. 11 (1), pp. 65–99.

Lawson, B. (1997) *How Designers Think - The Design Process Demystified*. Oxford: Architectural Press. Quoted in: Persson, S. and Anders, W. (2003) Relational Modes between Industrial Design and Engineering Design: A Conceptual Model for Interdisciplinary Design Work. *Proceedings of the 6th Asian Design International Conference*, Tsukuba.

Lueptow, R. M. (2000) *Graphics Concepts*. New Jersey: Prentice Hall.

Martin, J. (1989) *Technical Illustration*. London: McDonald and Co. In: Evans, M. A. (2002) *The Integration of Rapid Prototyping within Industrial Design Practice* (Staff Thesis). Loughborough: Department of Design and Technology Loughborough University.

Olofsson, E. and Sjölén, K. (2005) *Design Sketching*. Sundsvall, Sweden: Keeos Design Books AB.

Pavel, N. (2005) *The Industrial Designer's Guide to Sketching*. Trondheim: Tapir Academic Press.

Pipes, A. (2007) *Drawing for Designers*. London: Laurence King Publishing.

Powell, D. (1990) *Presentation Techniques*. London: Macdonald.

Raudebaugh, R. A. and Newcomer, J. (1999) *Visualization, Sketching and Freehand Drawing for Engineering Design*. Mission, Kansas: Schroff Development Corporation.

Robbins, E. (1994) *Why Architects Draw*. Cambridge, MA: MIT Press. Quoted in: Do, E. Y-L. (2002) Drawing Marks, Acts, and Reacts: Toward a Computational Sketching Interface for Architectural Design. *Artificial Intelligence for Engineering Design, Analysis and Manufacturing*, No. 16, pp. 149–171.

Song, S. and Agogino, A. S. (2004) Insights on Designers' Sketching Activities in Product Design Teams. In: *ASME Design Engineering Technical Conference '04* held in Salt Lake City, Utah. Quoted in: Yang, M. C. and Daniel, J. (2005) A Study of Prototypes, Design Activity, and Design Outcome. *Design Studies*, Vol. 26(6).

Tjalve, E. (1979) *A Short Course in Industrial Design*. London: Butterworth and Co.

Tjalve, E., Andreasen, M. M., et al. (1979) *Engineering Graphic Modelling: A Workbook for Design Engineers.* London: Butterworth and Co.

Tovey, M. (1989) Drawing and CAD in Industrial Design. *Design Studies,* Vol. 10(1), pp. 24–39.

Ullman, D. G., Wood, S., et al. (1990) The Importance of Drawing in the Mechanical Design Process. *Computer and Graphics,* Vol. 14(2), pp. 263–274.

Yang, M. C. (2003) Concept Generation and Sketching: Correlations with Design Outcome. In: *ASME Design Engineering Technical Conferences '03* held in Chicago, IL. Quoted in: Yang, M. C. and Daniel, J. (2005) A Study of Prototypes, Design Activity, and Design Outcome. *Design Studies,* Vol. 26(6).

6

Models

6.1 Models as design representation

According to Holmquist (2005), Models are non-functional objects used to describe the visual appearance of an intended design. However, Buur and Andreasen (1989b) indicate that they can also be used to reproduce the rough functional properties of a product. The term 'modelling' describes the creation and use of physical artefacts to 'elaborate, synthesise, evaluate and communicate' a design proposal (Andreasen 1994). Models are used because 2D Sketches and 2D Drawings are inadequate to explain the three-dimensional attributes of an object (Tovey 1997). As such, Models are physical objects that allow both industrial designers and engineering designers to explain and explore the function, performance and aesthetic aspects of a design, enabling them to 'describe, visualise and sculpture thoughts' (Buur and Andreasen 1989a) and to, 'develop, reflect, and communicate design ideas with others' (Peng 1994). Some examples of Models are shown in Figures 6.1 and 6.2. These models enable designers and engineers to produce variations of the design idea during the early and development stages of the design process.

However, Garner (2004) pointed out that some Models are more suitable for communicating information, while others are better suited for testing ideas. For example, the Appearance Model (Section 6.3) is often used during *Concept Design* and *Concept Development* phases to communicate overall form and/or aesthetic considerations in 3D. In contrast, a Conceptual of Operation Model (Section 6.5) is often employed to explore the potential of a particular functional component of a possible design solution. In either case, Models are used in practice as means to iteratively develop design ideas. To achieve this, Models may express intent at varying levels of fidelity across different dimensions of interest (i.e. form, technical functionality, product interaction, etc.). Figure 6.3 shows an Appearance Model and Figure 6.4 shows a Functional Concept Model.

Lucci and Oirlandini (1989) acknowledged that the translation from a 2D sketch or drawing into a 3D physical object is a significant phase of the design process. A full-size or scaled Model allows feedback from stakeholders, providing opportunity to iron out issues before committing to tooling or manufacture to minimise downstream mistakes (Powell 1990). Models as design representations are useful in illustrating how components are integrated or assembled so that clients may visualise a design (Woodtke 2000).

Brandt (2005) highlights the fact that Models function as boundary objects where each member, individual or stakeholder can converge towards a common understanding of the proposed design, yet remain in control of their professional interests. As boundary objects, the role and use of Models as design representations also indicate their distributed nature. Perhaps more so than Sketches (Chapter 4) and Drawings (Chapter 5), the physicality of representation through Models allow various

DOI: 10.1201/9781003227694-6

FIGURE 6.1
Example of a 3D Sketch Model (Choudhury, n.d.).

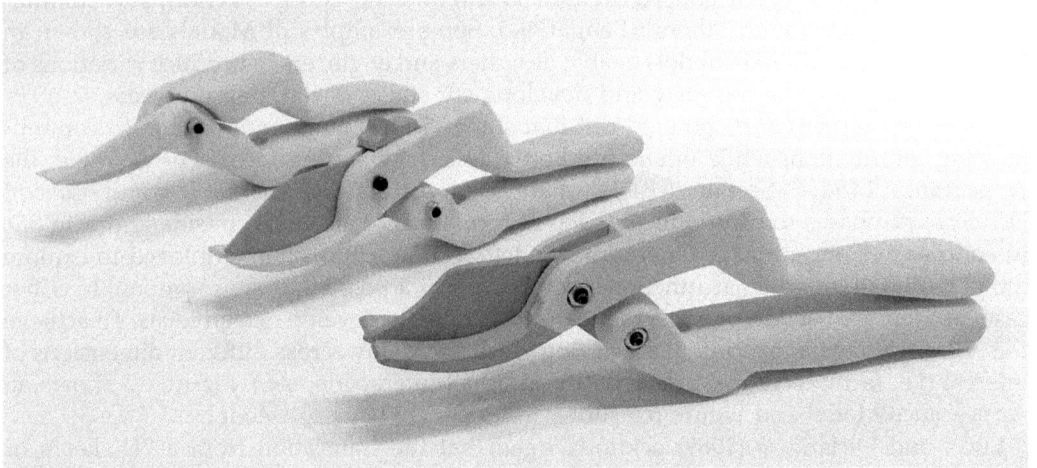

FIGURE 6.2
Functional Concept Models (Anderson, n.d.).

stakeholders the opportunity to evaluate design ideas. The potential of a design solution, as expressed in the model as a physical representation of intent, provides space
for individual responses, criticism and feedback that would otherwise not be possible
through 2D representation.

Baxter (1995) suggested that Models can be grouped into structural, functional, and structural and functional representations. Models can also vary according to the scale, accuracy and
material, and serve as an abstract, as well as a concrete representation to the final design

FIGURE 6.3
A Functional Concept Model (Choudhury, n.d.).

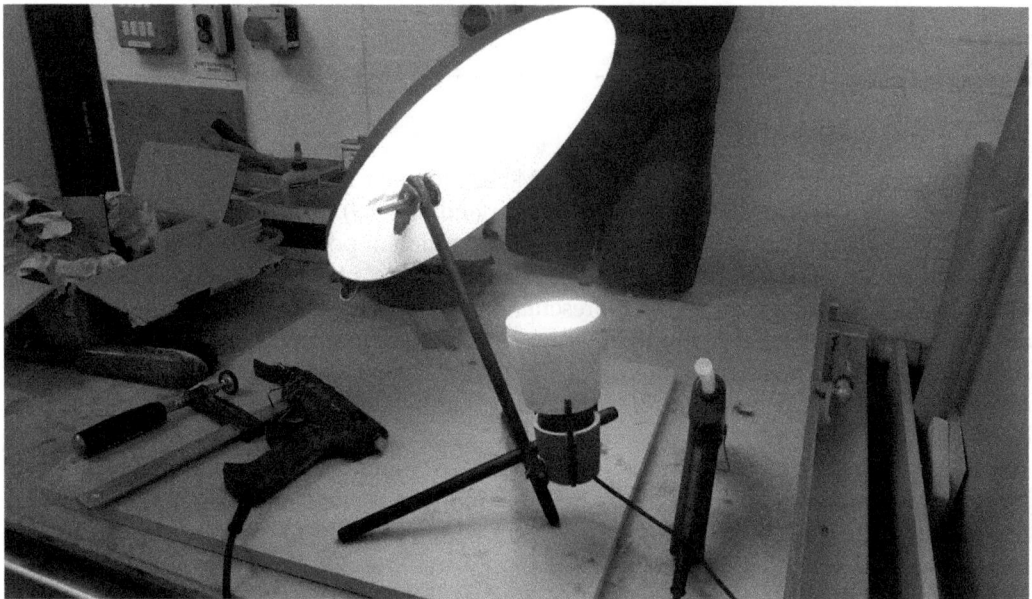

FIGURE 6.4
An Appearance Model (Choudhury, n.d.).

(Kvan and Thilakaratne 2003). They allow the developer to gain tactile clues (Ferguson 1992), described by Smyth (1998) as 'designers thinking with their hands', or a 'design-by-doing' activity described by Ehn and Kyng (1991). The act of modelling is comparable to Schön's (1983) description of a designer conversing with an image on paper. The sense of touch is important for perception and allows the developer to fully understand the form and geometry of the design. Models, like Sketches and Drawings, can act as a type of distributed cognition, where the Model provides opportunity for reflection on the potential of a solution idea, or part thereof. Unlike 2D representations, however, the physicality of the Model as it is worked in the hands, appears to provide a necessary multi-modality of intetaction with design representation, important to the design of physical products.

In terms of classification, Emori (1977) grouped Models as either qualitative or subjective. A qualitative Model emphasises aesthetics and is traditionally fabricated from solid materials since internal parts are unnecessary. In contrast, a subjective Model is more concerned with the functional aspects in terms of its performance and use. Simondetti (2002) claimed that building full-scale Models advance the understanding of a design by allowing physical demonstration of structural behavior as well as for visual presentation. Models like sketch representations, may be more or less ambiguous in their expression of design intent, dependent upon the aims of the Model in relation to the stage in progress. Unlike Sketches, Drawings and Illustrations, Models provide multi-model interaction (sight, touch) in their expression of design ideas. It appears the multimodal interaction with Models as design representations (i.e. visual, haptic) increase their effectiveness for communication of design intent between various stakeholders. Models provide enhanced communicative potential as design representations through their physicality.

6.2 3D Sketch Model

Also known as Sketch Models or 3D Rough Models (Garner 2006), a *3D Sketch Model* is used in a similar way to 2D sketching (Lucci and Oirlandini 1989). It is an affordable and quick way of physical representation that allows the exploration of potential ideas, obtaining visual and tactile feedback, and translating 2D representations into a tangible medium (Evans 2002). Often necessarily ambiguous in the representation of design ideas, due to a requirement for reflective interpretation related to the *Concept Design* phase in process, they are employed to support quick, 3D expression of intent. *3D Sketch Models* allow the exploration and evaluation of initial ideas. Like 2D sketching, *3D Sketch Models* also offer room for interpretation in their expression of intent.

Soft materials such as foam and balsa wood are used in their construction for achieving the general shape, and forming details with files, drills and sandpaper. These hand-processing techniques often provide designers with an ability to explore form as shapes are carven and refined through physical model making. Often expressing little or no technical functionality, *3D Sketch Models* primary role during concept design is to represent and communicate form factors (Figures 6.5 and 6.6). However, use-function aspects may also be considered through the 3D Sketch Models ability to express the physicality of form.

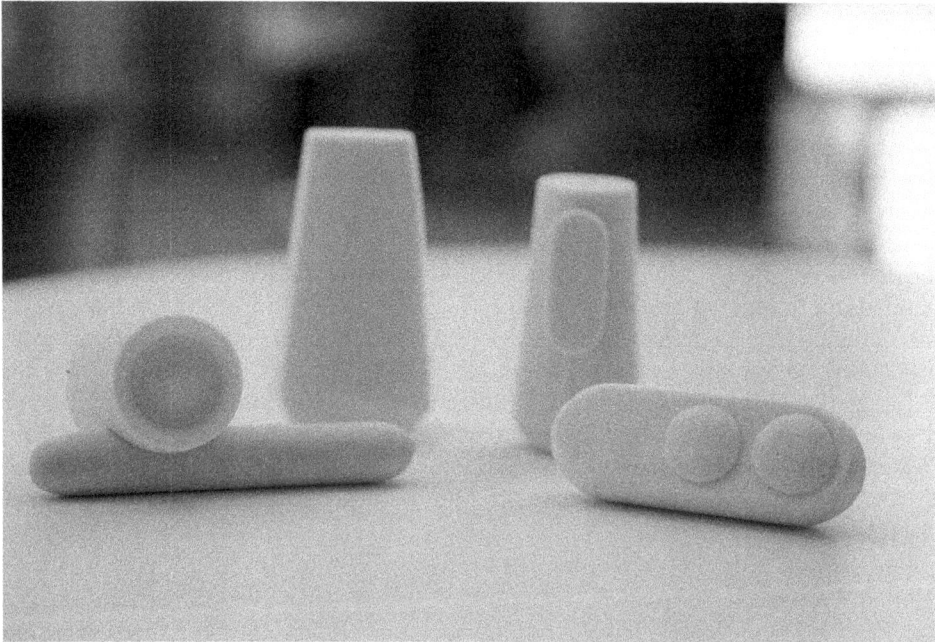

FIGURE 6.5
3D Sketch Models (Choudhury, n.d.).

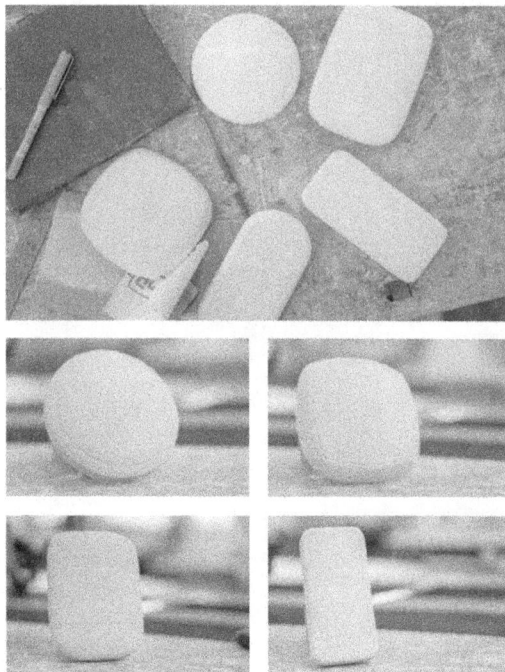

FIGURE 6.6
Other examples of 3D Sketch Models (Choudhury n.d.).

6.3 Design Development Model

Upon confirmation of a design concept, these Models are used to create a batch of accurate representations. They are employed to refine shapes, to investigate how components are fixed or for testing. They are created quickly with materials such as balsa wood and foam. To enhance realism, parting lines, slots and buttons may be drawn onto a material, as well as the use of paint and ready-made working parts. They often represent design intent at a higher degree of fidelity and with reduced ambiguity as ideas are elevated and a design direction starts to emerge and be refined.

Evans (1992) described these Models as Foam Models. However, as a wide range of materials may be applied, the term *Design Development Model* is used as a more inclusive term. *Design Development Models* are 3D design representations used to understand the relationships between components, cavities, interfaces, structure and forms (Figures 6.7 and 6.8). Often constructed at a higher level of fidelity compared to *3D Sketch Models* (Section 6.2), *Design Development Models* offer enhanced communication of various design details.

FIGURE 6.7
Design Development Models (Jameel, n.d.).

FIGURE 6.8
Design Development Models for VO Corporation and Nikon (Hill, 2011).

6.4 Appearance Model

The purpose of an *Appearance Model* is to enable stakeholders and clients to accurately evaluate the aesthetics of a design as compared to Sketches or Drawings (DTI 1992). *Appearance Models* are also known as Maquettes (Baxter 1995) or Block Models (Evans 1992). Powell (1990) suggests that these Models allow the design to materialise into a real physical form where for the first time stakeholders and clients are able to more holistically evaluate the design. However, it is important to note that they are only concerned with the external outlook without any functional features (Baxter 1995). As such, they are relatively unambiguous in the expression of product form and material elements on a visual dimension. However, functional aspects, as well as, material properties (i.e. weight, surface finishes, haptic perception) may be absent. This also relates to their purpose of use as means to communicate an aesthetic direction as a physical representation, rather than a functional Prototype (Chapter 7).

In terms of fabrication, a wide variety of materials may be used, including wood, plastics, metal, fibreglass, etc. The model is usually finished to a high level of surface treatment and complete with decals to closely resemble the final product. It is close in its fidelity to the potential manufactured product in terms of visual aspects. Increasingly, rapid prototyping technologies have enabled detailed parts to be fabricated, shortening the model-making time. Simply put, *Appearance Models* are 3D visual design representations that realistically define the visual aspects of a product, but do not contain any or limited working mechanisms (Figures 6.9 and 6.10).

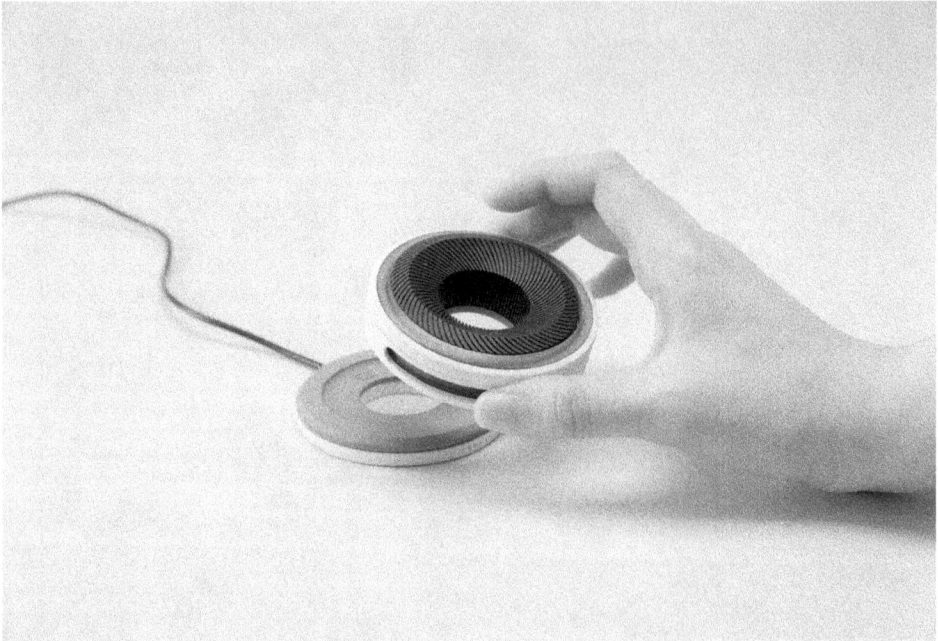

FIGURE 6.9
An Appearance Model (Choudhury, n.d.).

FIGURE 6.10
Close-up of an Appearance Model (Choudhury, n.d.).

6.5 Functional Concept Model

Functional Concept Models are used to investigate the working parts of a product concerning aspects such as yield and performance (Buur and Andreasen 1989b). They are also known as Principle Models (Evans 1992) or Principle-Proving Models (Garner 2006) to prove that a technology or a functional part works. They are mechanical-looking and do not have the appearance of the final product. *Functional Concept Models* are 3D visual design representations that show the functionality and highlight important functional parameters including yield and performance factors.

Functional Concept Models may also be described as Works-like Models, in their focus on technical functionality and use. They are employed during *Concept Design* or *Concept Development* phases of the design process to test, explore and validate potential solutions along the dimension of functionality. In this role, they are often employed as Proof-of-Concept Models or Proof-of-Principle Models where the purpose of representation is to test, refine and confirm aspects related to technical functionality and use (Figures 6.11 and 6.12). They may attempt to approximate function as far as possible (fidelity) in order to be best placed to provide valid feedback and opportunities for incremental, functional improvements.

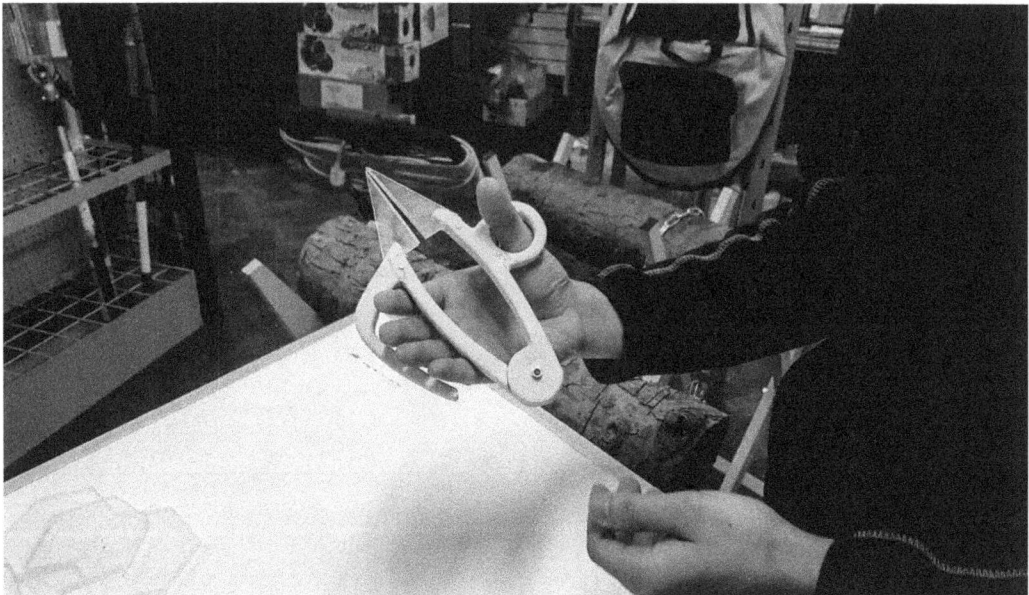

FIGURE 6.11
A Functional Concept Model (Choudhury, n.d.).

FIGURE 6.12
Another example of a Functional Concept Model (Anderson, n.d.).

6.6 Concept of Operation Model

According to Buur and Andreasen (1989b), these Models are used to show how the product would be operated, controlled or managed. *Concept of Operation Models* are 3D visual design representations that help communicate operational strategies and usage procedures relating to the product. They aim to communicate how a product may locate within and be used in context, often by various stakeholders (Figure 6.13).

FIGURE 6.13
A Concept of Operation Model (Aligizakis, n.d.).

6.7 Production Concept Model

The term production refers to the process of how things are made, produced or manufactured (Longman Dictionary 2005). *Production Concept Models*, therefore, describe a type of physical representation that allows the product developer and other stakeholders to understand, evaluate and prepare the design for production (Buur and Andreasen 1989b). *Production Concept Models* allow the assessment of processes, costs and requirements before committing to manufacture proper. They aim to provide opportunity for evaluation and confirmation of approaches to final manufacture. Often unambiguous in their expression of detailed intent, *Production Concept Models* allow validation of approach to production, as well as incremental refinement of production approach. Through their physicality, they also support increased interrogation of fit, moulding and potential production processes. Thus, *Production Concept Models* are 3D visual design representations used to help assist the evaluation of production processes or manufacturing technologies for final production (Figures 6.14–6.16).

FIGURE 6.14
A Production Concept Model rendered in 3D CAD (Choudhury, n.d.).

FIGURE 6.15
The actual Production Concept Model (Choudhury, n.d.).

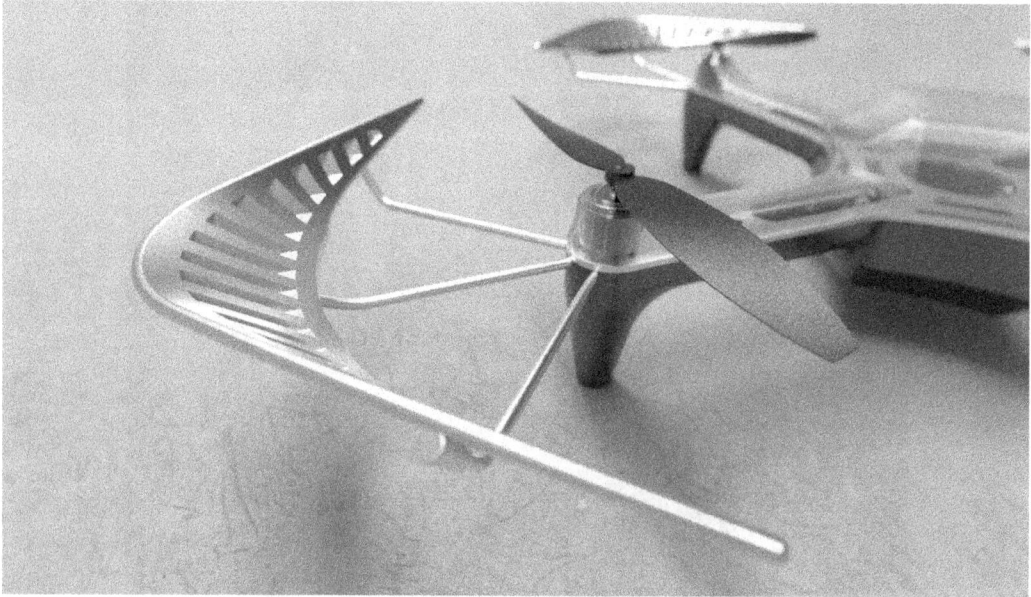

FIGURE 6.16
Detail of a Production Concept Model (Choudhury, n.d.).

6.8 Assembly Concept Model

Assembly refers to fitting or putting parts together (The Merriam-Webster Dictionary 1994). These physical models allow developers to establish and ascertain aspects concerning the assembly of a product. They allow issues relating to costs and investments in equipment to be evaluated early in the development stages (Buur and Andreasen 1989b).

Assembly Concept Models are 3D design representations that provide information regarding the component relationships in terms of assembly, cost and investment. As such they are often produced at a high degree of fidelity, closer to a propositional design outcome. They are often employed to help validate the final detail of a design (Section 6.7), however, *Assembly Concept Models* focus upon the assembly of part and product architecture during a process of part manufacture and product production (Figures 6.17–6.19).

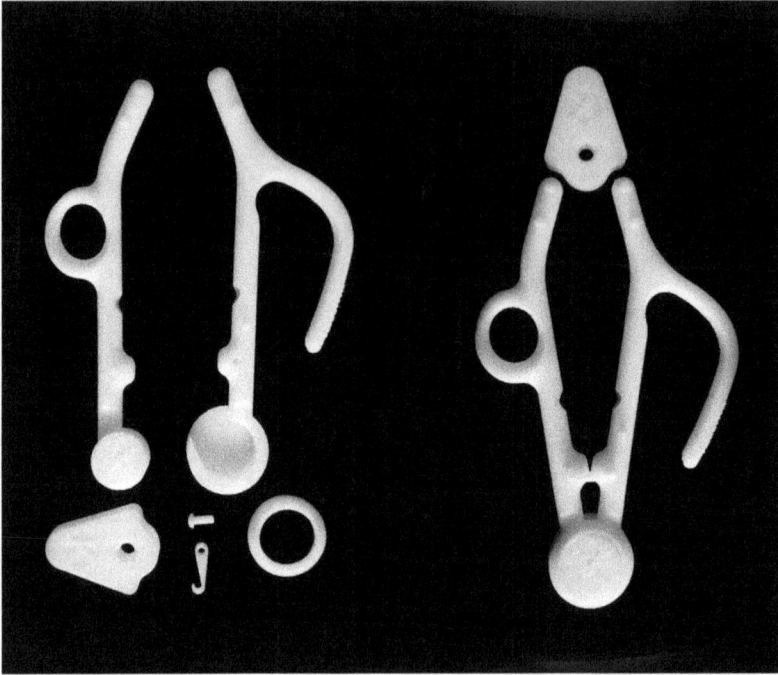

FIGURE 6.17
An Assembly Concept Model (Choudhury, n.d.).

FIGURE 6.18
An example of an Assembly Concept Model (Man, n.d.).

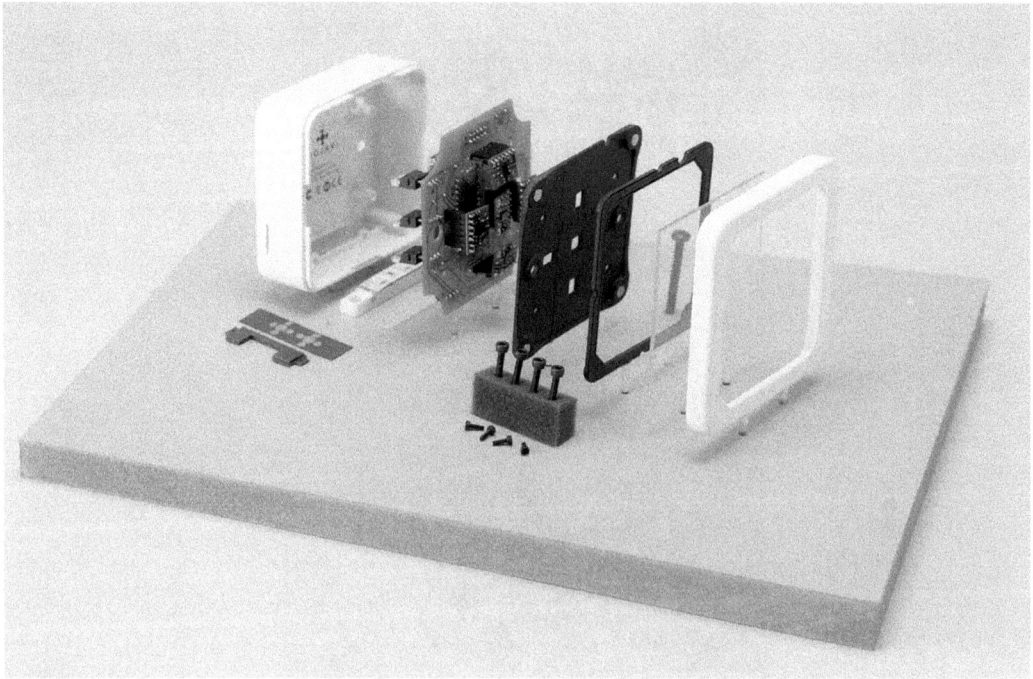

FIGURE 6.19
Another example of an Assembly Concept Model (Mortimer, n.d.).

6.9 Service Concept Model

Service Concept Models are 3D design representations that help illustrate how a design may be serviced or maintained. According to Buur and Andreasen (1989b), during the development process, it is important to consider how the product could be cleaned and serviced throughout its lifecycle. These Models help developers to establish solutions such as how a user would install a new set of batteries, or how a service technician would be able to disassemble a product for repairs.

Service Concept Models are also often used as stimuli in-user studies to explore possibilities for improvement and opportunities for development. *Service Concept Models* aim to represent a particular functional aspect of a design to a high degree of fidelity in order to validate and or improve on a functional requirement (Figures 6.20 and 6.21). In this, the focus may be upon a particular function and user interaction, in terms of servicing or product maintenance (Figures 6.22 and 6.23).

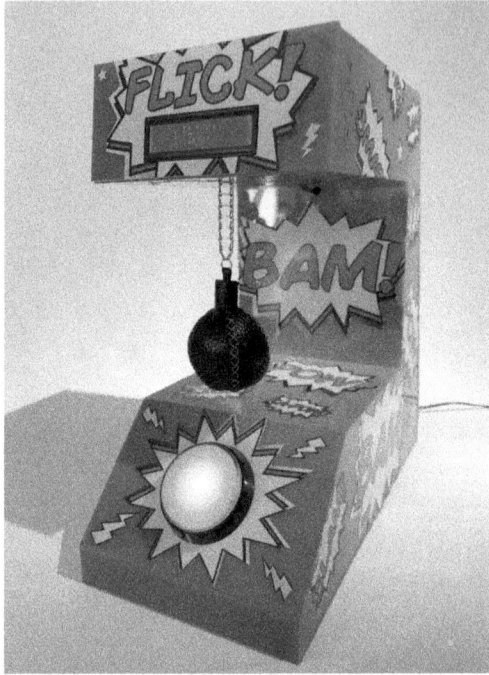

FIGURE 6.20
A Service Concept Model (Man, n.d.).

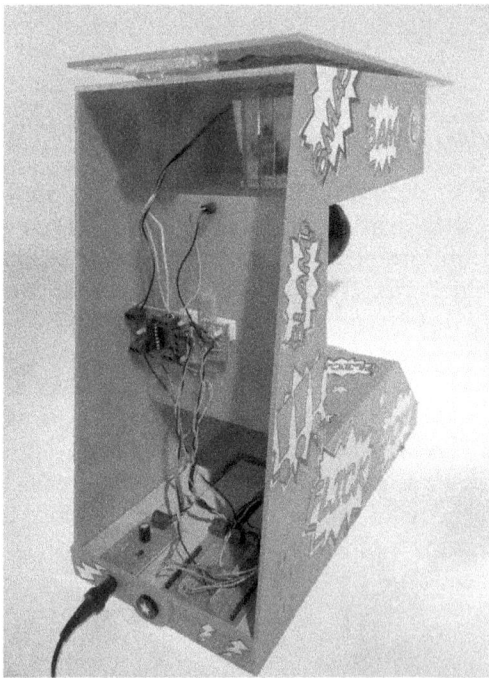

FIGURE 6.21
Rear view of a Service Concept Model (Man, n.d.).

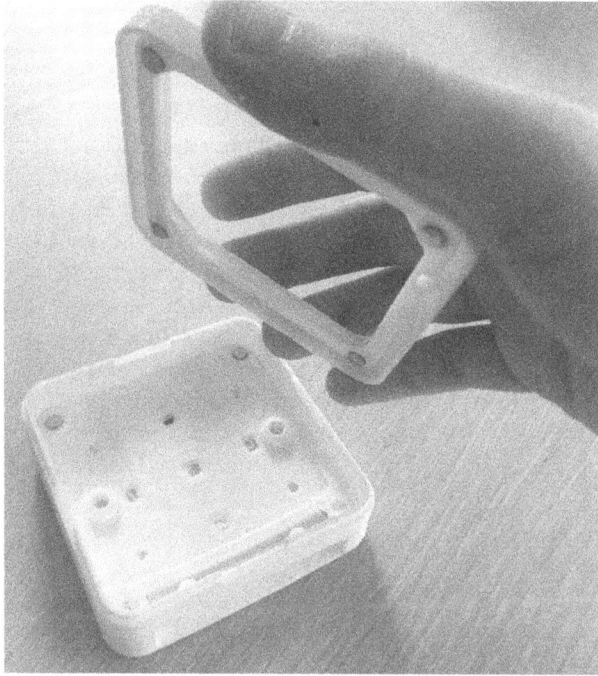

FIGURE 6.22
Another example of a Service Concept Model (Mortimer, n.d.).

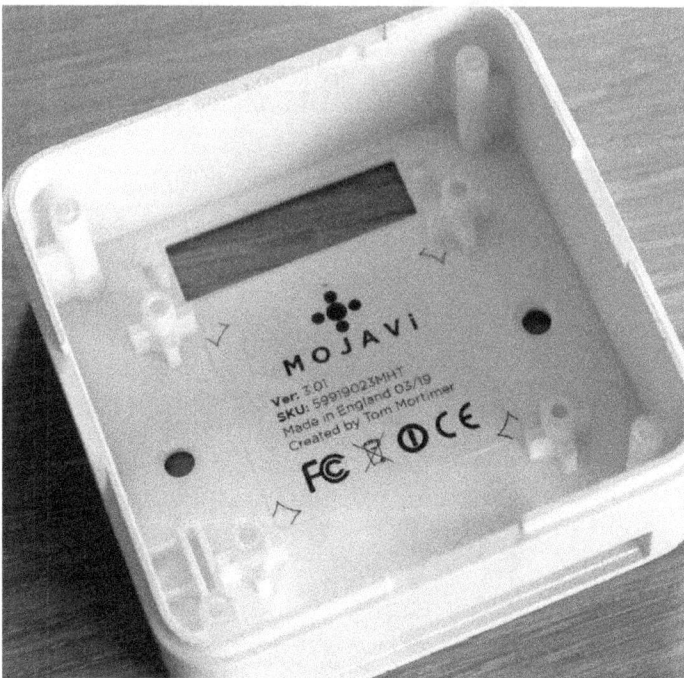

FIGURE 6.23
A close-up view of a Service Concept Model (Mortimer, n.d.).

6.10 Summary

In this chapter, we have explored the role and use of Models as representations of design intent. An introductory section (Section 6.1) outlined the physical character of Models as design representations against the 2D representations of sketching (Chapter 4) and drawing (Chapter 5). We have discussed the physicality of Models in terms of opportunity for enhanced communication between various stakeholders at different phases of the design process. Section 6.1 discussed how physicality in the expression of intent may support enhanced understanding as model representations act as boundary objects between individuals. The various uses of Models as design representations was further discussed in terms of different purposes for communication, exploration and design development. The discussion extended to existing attempts to classify various types of Models through a criteria that describe their purpose of use. Figures 6.24, 6.25 and 6.26 show examples of the PUNKT Phone by Jasper Morrison as a semi functional appearance model.

FIGURE 6.24
Stage during the final assembly of Pre-Production Prototype casing. (Hill, 2011. For Jasper Morrison Studio).

FIGURE 6.25
Key Caps for Reference model (Hill,2011 for Jasper Morrison Studio).

FIGURE 6.26
Fully Assembled Reference Model (Hill, 2011.).

A taxonomic classification of eight Models have been presented (Sections 6.2–6.9). For each, the Model type was discussed in relation to its application in support of design practice, the phase in design process it may be used (i.e. *Concept Design, Concept Development, Detail Design*), and the processes and materials used in its construction. The eight cited Models were also described in terms of relations between the type of Model used, phase in design process, communication and design development requirements (Figures 6.27 and 6.28). As with 2D

FIGURE 6.27
Early ID Concept Model in Foam describes physical volume, form and ergonomics. (Hill, 2000; Samsung Design Europe Studio).

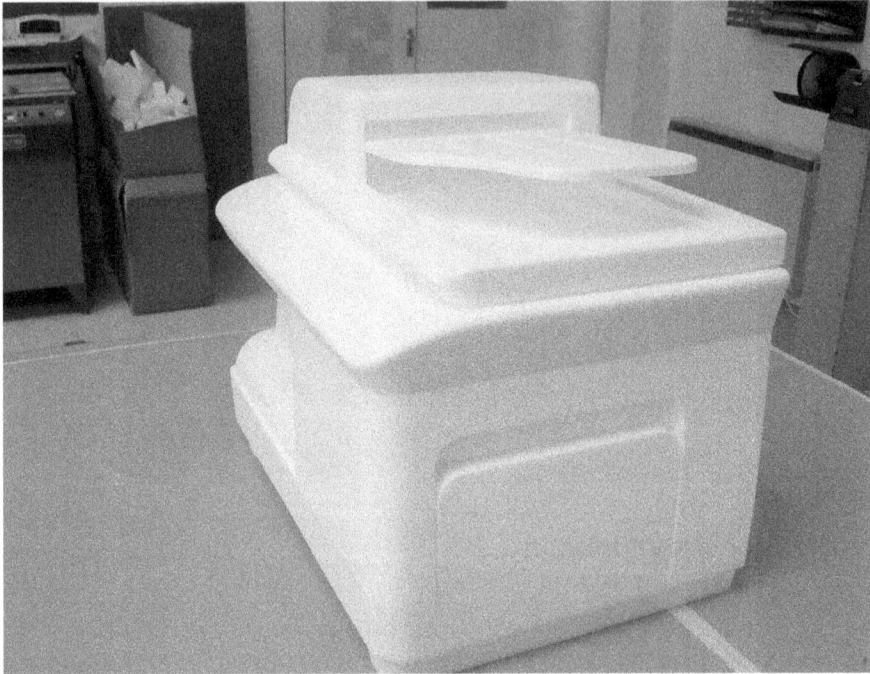

FIGURE 6.28
Early ID Concept Model in Foam describes physical volume, form and ergonomics. (Hill, 2000; Samsung Design Europe Studio).

design representation (Chapters 4: Sketches, and Chapter 5: Drawings), we do not claim that our taxonomy of Models is inclusive and complete. Instead we provide a distinction between eight types of model representations as means to explore the various uses of Models, and their role as representations of design intent on the various dimensions of their use.

References

Andreasen, M. M. (1994) Modelling - The Language of the Designer. *Journal of Engineering Design*, Vol. 5(2), pp. 103–115.

Baxter, M. (1995) *Product Design: A Practical Guide to Systematic Methods of New Product Development*. London: Chapman and Hall.

Brandt, E. (2005) How do Tangible Mock-Ups Support Design Collaboration? *Proceedings of the Nordic Design Research Conference, 'In the Making'*, Copenhagen, Denmark.

Buur, J. and Andreasen M. M. (1989a) Design Models in Mechatronic Product Development. *Design Studies*, Vol. 10(3), pp. 155–162.

Buur, J. and Andreasen, M. M. (1989b) Design Models in Mechatronic Product Development. *Design Studies*, Vol. 3, pp. 105–162. Quoted in: Andreasen, M. M. (1994) Modelling - The Language of the Designer. *Journal of Engineering Design*, Vol. 5(2).

DTI (1992) *Managing Product Design Projects*. London: Department for Trade and Industry.

Ehn, P. and Kyng, M. (1991) Cardboard Computers: Mocking It-Up or Hands-On the Future. In: Greenbaum, J. and Kyng, M. (1992) *Design at Work: Cooperative Design of Computer Systems*.

Mahwah, NJ: Lawrence Erlbaum pp. 169–195. Quoted in: Eva B. (2005) How Tangible Mock-Ups Support Design Collaboration. *Proceedings of the Nordic Design Research Conference, 'In the Making'* Copenhagen, Denmark.

Emori, R. I. (1977) *Scale Models in Engineering*. London: Pergamon Press. In: *Engineering Designer* (January 1999) Wiltshire: Institution of Engineering Designers. In: Evans, M. A. (2002) *The Integration of Rapid Prototyping within Industrial Design Practice*. Staff Thesis. Loughborough: Department of Design and Technology, Loughborough University.

Evans, M. (1992) Model or Prototype Which, When and Why? *IDATER 1992 Conference*, Loughborough University (Design and Technology).

Evans, M. A. (2002) *The Integration of Rapid Prototyping within Industrial Design Practice*. Staff Thesis. Loughborough: Department of Design and Technology, Loughborough University.

Ferguson, E. S. (1992) *Engineering and the Mind's Eye*. Cambridge, MA: MIT Press.

Garner, S. (2004) *T211 Design and Designing: An Introduction to Design and Designing*. Milton Keynes: The Open University Press.

Garner, S. (2006) *Modelling Workbook 1: T211 Design and Designing Workbook 1 Technology Level 2* (2nd ed.). Milton Keynes: The Open University Press.

Holmquist, L. E. (2005) Prototyping: Generating Ideas or Cargo Cult Designs? *Interactions*, Vol. 12(2), pp. 48–54.

Kvan, T. and Thilakaratne, R. (2003) Models in the Design Conversation: Architecture vs Engineering. *Proceedings of the AASA Conference*. Melbourne: Association of Architecture Schools of Australasia.

Longman Dictionary of Contemporary English (2005) D. S. (Ed.). Essex: Pearson Education Limited.

Lucci, R. and Oirlandini, P. (1989) *Product Design Models*. New York: Van Nostrand Reinhold. In: Evans, M. A. (2002) *The Integration of Rapid Prototyping within Industrial Design Practice*. Staff Thesis. Loughborough: Department of Design and Technology Loughborough University.

Peng, C. (1994) Exploring Communication in Collaborative Design: Co-Operative Architectural Modelling. *Design Studies*, Vol. 15(1), pp. 19–44.

Powell, D. (1990) *Presentation Techniques*. London: Macdonald.

Schön, D. A. (1983) *The Reflective Practitioner: How Professionals Think in Action*. London: Temple Smith.

Simondetti, A. (2002) Computer Generated Physical Modeling in the Early Stages of the Design Process. *Automation in Construction*, Vol. 11, pp. 303–311.

Smyth, M. (1998) The Tools Designers Use: What Do They Reveal about Design Thinking? *Idater 98*. Loughborough University.

The Merriam-Webster Dictionary (1994) F. C. Mish (Ed.). Massachusetts: Merriam-Webster.

Tovey, M. (1997) Styling and Design: Intuition and Analysis in Industrial Design. *Design Studies*, Vol. 18(1), pp. 5–31.

Woodtke, M. V. (2000) *Design with Digital Tools*. Blacklick: McGraw Hill, pp. 108. In: Evans, M. A. (2002) *The Integration of Rapid Prototyping within Industrial Design Practice*. Staff Thesis. Loughborough: Department of Design and Technology, Loughborough University.

7

Prototypes

7.1 Prototypical design representation

The aim of prototyping is to confirm the results of the design process and the decisions that have been made towards the development and confirmation of design intent. Prototypes are used to communicate the final design (Kurvinen et al. 2008). In the context of design, there are several definitions for the term *Prototype*. According to Holmquist (2005), Prototypes consist of functional parts and do not resemble a final product. Other researchers clarified that Prototypes are full-scale physical representations (Luzadder 1975; Evans 1992), while Best (2006) suggested that Prototypes can be either physical or virtual. Other related words such as *Rapid Prototyping* also refer to the terms *Additive Manufacturing* or *3D Printing*, while *Virtual Prototyping* refers to digital representations created through computer simulation within an augmented environment (i.e. virtual reality), or hybrid physical/digital representations (augmented reality). Whatever the media of expression, Prototypes aim to approximate a solution on various dimensions (dependent upon the purpose of use), and its varying levels of detail and fidelity. In this sense, Prototypes refer to a tangible example of a potential solution used as means to offer opportunities for communication, design refinement and development. For the purposes of our discussion of design representation, we define *Prototypes* as full-scale 3D visual design representations that incorporate working components.

Kelly and Littman (2001) described Prototypes as being 'worth a thousand pictures'. They serve as a tangible artefact providing confidence to stakeholders about the final design (Kelley 2001). With a physical representation, stakeholders can interact and finalise aspects of the design (Bødker and Buur 2002; Preece et al. 2002). They bring multi-disciplinary perspectives together and act as a medium where joint decisions can be made and for refinements to be conducted safely and cheaply (Kolodner and Wills 1996). According to Subrahmanian et al. (2003), Prototypes are not static and they may dynamically develop as the design further progresses. Otto and Wood (2001) clarify that multidisciplinary stakeholders use Prototypes differently according to their needs.

For example, industrial designers use Prototypes to investigate the look and feel of a design, while mechanical engineers utilise Prototypes to analyse functional properties. As a physical working representation of a design proposal, Prototypes are used to test the feasibility of the finalised concept, for customer feedback and to clarify production and technical issues (Figures 7.1–7.3) (Holbrook and Moore 1981; Finn 1985). Prototypes may often sit on a continuum between ambiguity and clarity of representation, dependent upon their purpose of use. Likewise, the degree of fidelity can depend upon what aspect of a design is being represented through a Prototype and for what purpose according to the design process.

DOI: 10.1201/9781003227694-7

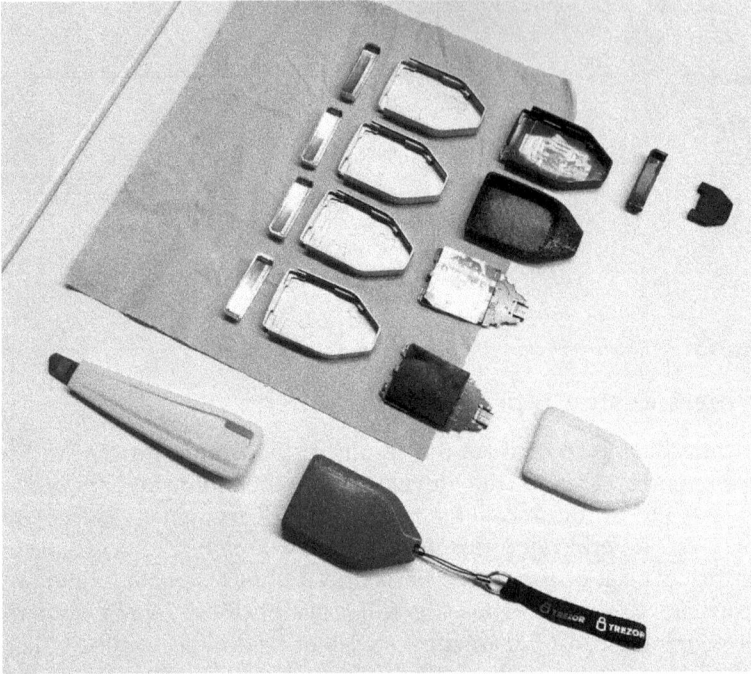

FIGURE 7.1
Prototype components, pre- and post-assembly (Prototypum, n.d.).

FIGURE 7.2
Close-up view of a Prototype showing the material and finish (Prototypum, n.d.).

FIGURE 7.3
A close-up view of another Prototype (Prototypum, n.d.).

Yang and Daniel (2005) suggest that the process of constructing Prototypes itself allows developers to understand issues first-hand that cannot be gained from 2D Drawings or computer Models alone. An example is the plywood chair built by Morrison (1990) where a hands-on approach in the construction enabled the designer to develop a deeper understanding of the design and support explanation of intent to manufacturers. This means that Prototypes require greater commitment in terms of skill, time and cost as compared to other representations. They may be created in a specialist in-house workshop or outsourced to an external contractor (Avrahami and Hudson 2002). It is also important for the Prototype to closely resemble the actual product to avoid false expectations (Rosenberg 2006). Models are better suited during the early stages of development for problem-solving and idea generation (i.e. *Concept Design*), whereas Prototypes are employed towards the later stages (*Development Design* and *Detail Design*) to confirm and evaluate the aesthetics, ergonomics and performance of the design. Prototyping earlier in the design cycle rather than later is also linked to better results as it allows for more time, iteration and refinements to be made (Yang 2009; Ward et al. 1995).

As an integration medium, Prototypes show how components may fit together and are used to detect discrepancies. In terms of project milestones, Prototype representations can demonstrate an agreed level of progress has been acheived. Prototypes are also used by manufacturers to confirm tooling, for cost analysis and as promotional material. In addition, Ulrich and Eppinger (2003) noted that products with high technical or market risk tend to require more Prototypes to be built and tested. Evans (1992) pointed out that it is important to understand the underlying reason for producing a Prototype so that the correct design intention can be communicated. For example, a Functional Prototype may look unattractive, but its purpose is to illustrate the mechanical aspects and not its aesthetic content (i.e. form and CMF design) (Figure 7.4).

The various purposes Prototypes are employed to serve, then, also relates to the notion of dimension of interest, where representational fidelity can be high on one dimension (i.e. a functional aspect of a proposed design), but lower or non-existent on another (i.e. expression of form, colour and material choices). A Prototype may contain several uses. For instance, a Proof-of-Concept Prototype showing functional aspects may also be useful for developers to examine its mechanism, size and dimensions. In classifying Prototypes, Sommerville (1995) grouped Prototypes as throwaway, evolutionary, or incremental.

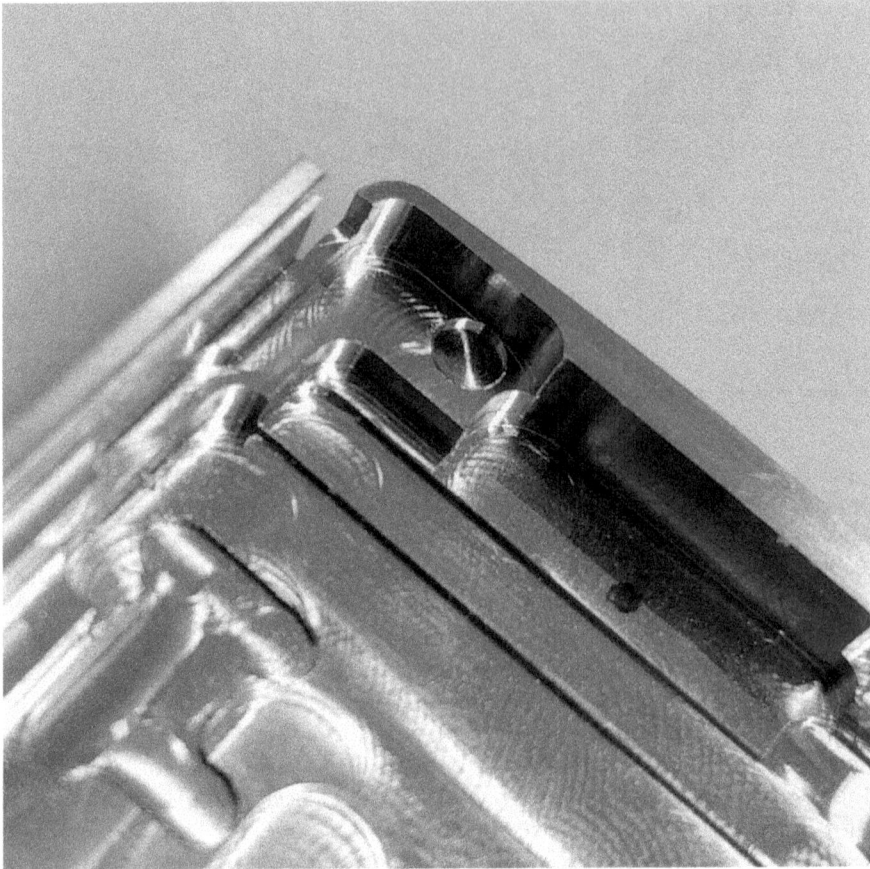

FIGURE 7.4
An internal view of a Prototype showing machined parts (Prototypum, n.d.).

A Throwaway Prototype is often used early in the development stage for clarifying ideas. These Prototypes often lack specific detail and are ambiguous in their representation of intent. Evolutionary Prototypes are continually developed and evaluated; while Incremental Prototypes bring small changes to the design (Figures 7.5 and 7.6). This classification is also similar to that of Budde et al. (1992) who classified Prototypes as evolutionary, experimental,

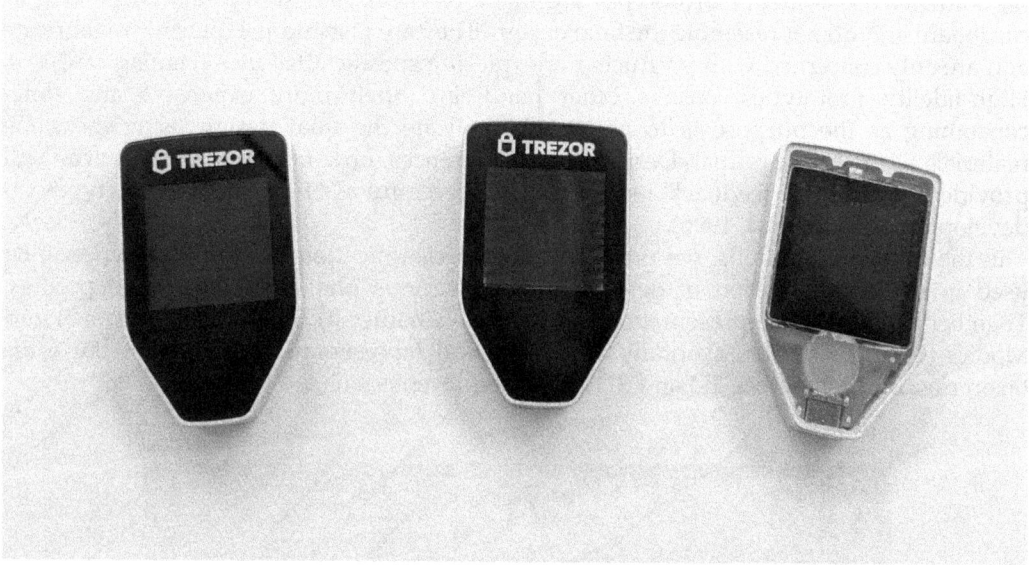

FIGURE 7.5
Final Hardware Prototypes (Prototypum, n.d.).

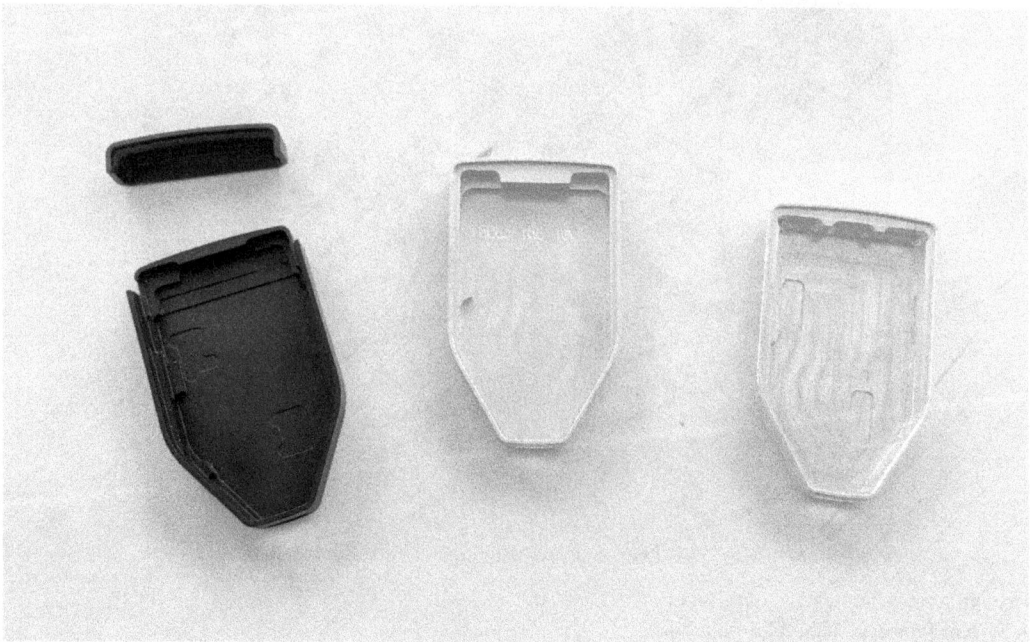

FIGURE 7.6
Internal view of the Final Hardware Prototypes (Prototypum, n.d.).

and exploratory. In another classification, Ulrich and Eppinger (2003) grouped Prototypes according to their degree of comprehensiveness.

A Comprehensive Prototype is described as a full-scale working version of the product shown to clients and potential customers to evaluate the overall design. Focused Prototypes, on the other hand, contain some characteristics such as having only electronic parts (Figure 7.7). Likewise, Preece et al. (2002) classified Prototypes as low-fidelity or high-fidelity. Low-fidelity Prototypes are made of cheap and simple materials such as cardboard and do not resemble the final design. They are fast and inexpensive to fabricate and are only concerned with producing or exploring specific attributes (Hanington 2006). High-fidelity Prototypes, on the other hand, are often more expensive and time-consuming as the purpose is to accurately replicate the final design using the same materials as the final product. Despite these differences, Low-fidelity Prototypes can still provide the necessary feedback and are just as successful as High-fidelity Prototypes for development (Virzi et al. 1996).

In the following sections, we provide our own classification of nine Prototypes often used in the representation of design intent as various phases of the design process. Together with design representation as Sketches (Chapter 4), Drawing (Chapter 5) and Models (Chapter 6), our taxonomy of prototypical representations completes our four-taxon classification across 2D and 3D design representations.

FIGURE 7.7
A System Prototype (Choudhury, n.d.).

7.2 Appearance Prototype

According to Evans (2002), *Appearance Prototypes* define the physical outlook of a design idea, as well as integrating the functional components. They have also been termed Integration Prototypes (Yang and Daniel 2005). Knoblaugh (1958) emphasised that *Appearance Prototypes* resemble the production item and are a check before tooling. In this sense, they often express design intent on the form aesthetic dimension at a higher level of fidelity and without ambiguity. Findings by Evans and Campbell (2003) provided evidence showing *Appearance Prototypes* have been useful in helping developers to evaluate the final design and the user interface prior to manufacturing. In distinguishing an Appearance Model and an *Appearance Prototype*, the latter is more complicated as it integrates function and aesthetics; whereas an Appearance Model only defines the exterior surface with no internal components. Due to the high level of detail and cost, *Appearance Prototypes* are usually made during the final stages of development. Additive Manufacturing is increasingly used to fabricate components for Appearance Prototypes as it allows complex and delicate parts to be made that are not possible to be created by hand. In line with Evans (1992), Appearance Prototypes are highly detailed, full-scale 3D visual design representations that combine function and aesthetics (Figures 7.8 and 7.9).

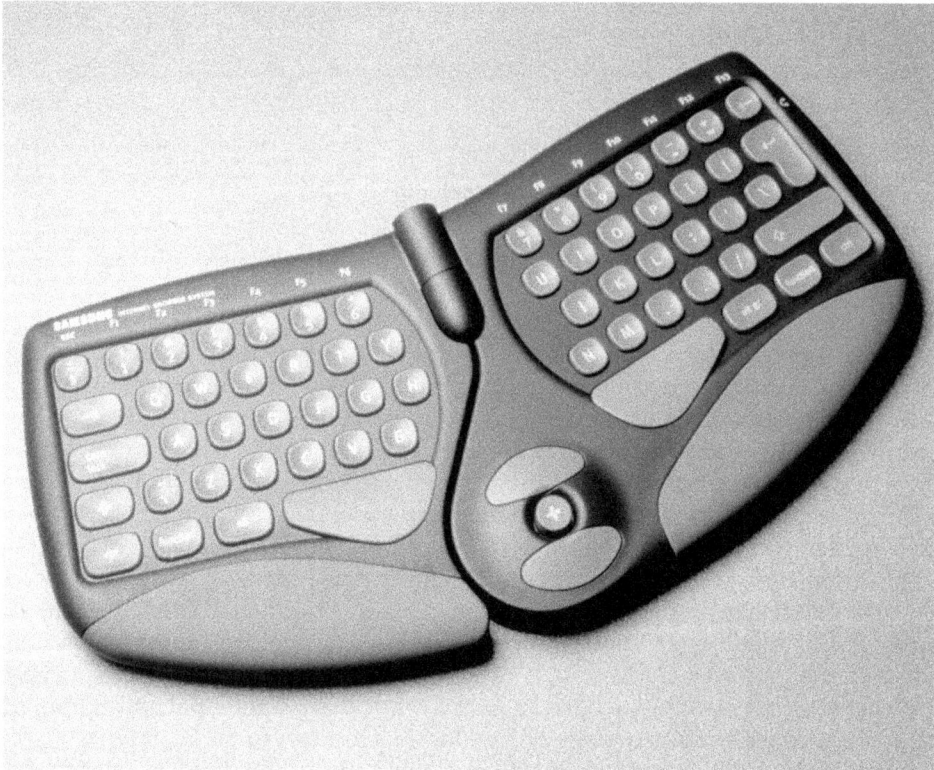

FIGURE 7.8
Appearance Prototype of folding keyboard. Future portability concept informed by developing technologies (Hill, 1997; Samsung Design Europe Studio).

FIGURE 7.9
Another example of an Appearance Model, 1 of 3 studies for consumer evaluation (Hill, 2001; for Samsung Design Europe Studio).

7.3 Alpha Prototype

Also known as First Prototypes (Veveris 1994), the *Alpha Prototype* incorporates the material and layout that would be used for the actual product. They bring together parts that have been proven and fabricated with the same grade of materials as the actual product. However, the parts are produced in low volume using techniques such as rubber moulding instead of injection moulding. According to Ulrich and Eppinger (2003), *Alpha Prototypes* are mainly used by industrial designers to verify the outlook, or sometimes by engineering designers for strength and impact tests. "Alpha Prototypes" should not be confused with "Off-Tool Prototypes" discussed in Section 7.10. Of-Tool Prototypes use actual production methods and usually also the actual materials which is shown in Figure 7.10 and have much greater detail.

Design representations as *Alpha Prototypes* may communicate intent on various dimensions, but at a lower level of detail and fidelity compared to Beta Prototypes (see below). *Appearance Prototypes* are often used to explore design potential and communicate intent at a Concept Developed stage of the design process. Alpha Prototypes as design representations are 3D expressions of design intent also often used to verify the construction of sub-systems that have been individually proven and accepted with the materials, aesthetics and layout for the actual product such as those produced by Cambridge Industrial Design (Figures 7.11). They are important to the validation and further development of design features, offering some detail of intent, but at a level of fidelity to provide space for alternative possibilities and revision of solution ideas.

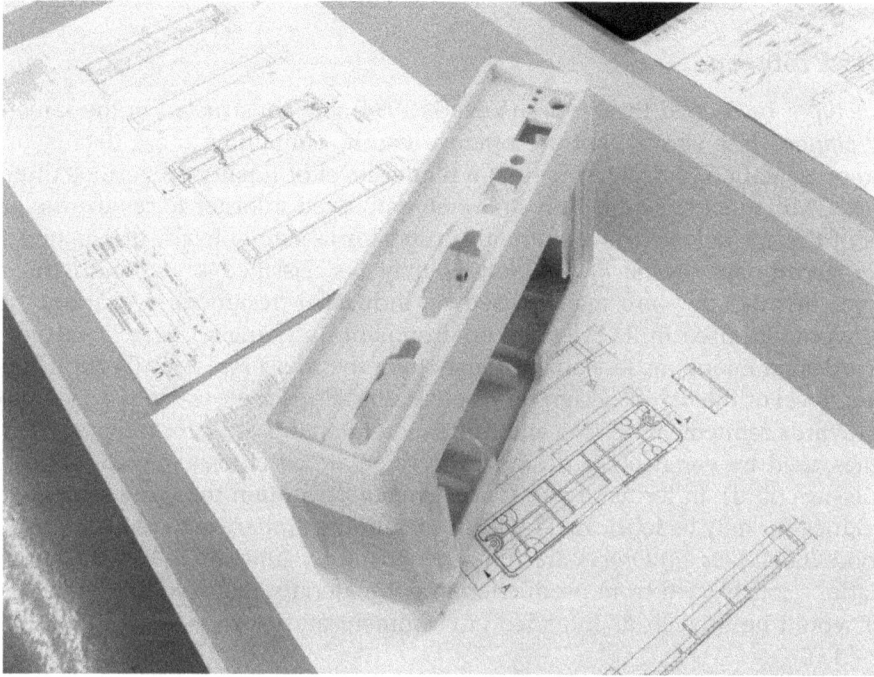

FIGURE 7.10
An Off-Tool Prototype with Technical Drawings on the work desk (Hill, 2011; For Cambridge Industrial Design and Accutronics).

FIGURE 7.11
Earlier stage of Alpha Prototype (Hill, 2011.; For Cambridge Industrial Design and Accutronics).

7.4 Beta Prototype

Beta Prototypes, or Second Prototypes (Veveris 1994), are constructed in the same way as *Alpha Prototypes* (see above), but are often full-scale, containing more details, and providing representation of design intent at a higher level of fidelity on various dimensions (i.e. form, CMF, function, etc.). They are often employed in order to review the resolved features of the *Alpha Prototype* and are also used for assembly trials, production evaluation and performance tests. *Beta Prototypes* may be classified under industrial design Prototypes because they are mainly used by industrial designers to examine how the product would be used in its intended environment. In doing so, *Beta Prototypes* aim to represent design intentions less ambiguously than the *Alpha Prototype* (Section 7.2). Their increasing level of fidelity of design representation reflects their use in practice, as design moves towards refinement, testing and more incremental change. *Beta Prototypes* are also sometimes used by engineering designers to calculate final costs and to work out regulatory issues (ibid). In terms of parts, *Beta Prototypes* contain the same materials as the final product but may be fabricated by CNC machining and are assembled by hand (Otto and Wood 2001). *Beta Prototypes* are full-scale and fully functional 3D visual design representations constructed from production intent materials and used to examine how the product would be used in its intended environment and to work out regulatory issues (Figure 7.12).

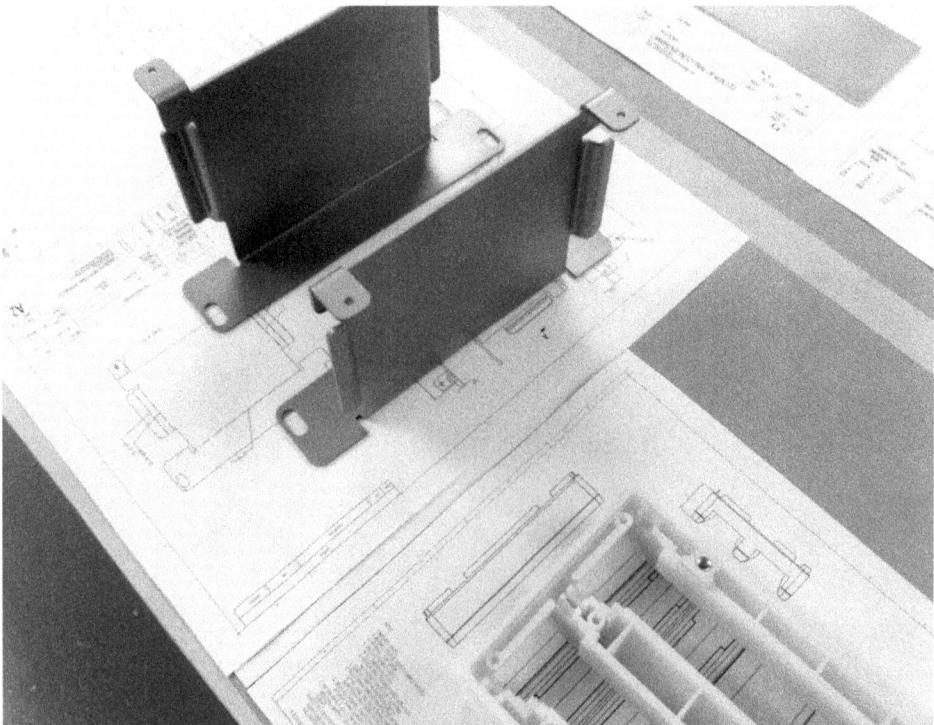

FIGURE 7.12
Beta Prototypes (Hill, 2011; For Cambridge Industrial Design & Accutronics).

7.5 Pre-Production Prototype

Pre-Production Prototypes, Pilot-production Prototypes (Ulrich and Eppinger 2003), Third Prototypes or Final Prototypes (Veveris 1994) are the final class of 3D design representations where all issues have been worked out and the design is ready for tooling and production (Otto and Wood 2001). At this stage, the production line is ready for a pilot-run and a short production run is undertaken to verify quality in terms of assembly and finish (Evans 1992). They therefore represent design intent at a very high level of fidelity, often being indistinguishable from a final manufactured product in both form and function. Pre-Production Prototypes are classed as industrial design Prototypes as most of the engineering details are fully resolved, they are therefore used to define the final look & function for final production (Figures 7.13 and 7.14). They are unambiguous in this objective, allowing complete communication of a final design solution. *Pre-Production Prototypes* are also used to gauge the manufacturing capability, with the parts sent to the clients for feedback. Consequently, Pre-production Prototypes are final 3D design representations used to check the product and its finishing as a whole and to perform production and assembly assessment in small batches.

FIGURE 7.13
A Pre-Production Prototype (Hill, 2011; For Cambridge Industrial Design & Accutronics).

FIGURE 7.14
Components of a Pre-Production Prototype (Hill, 2011; For Cambridge Industrial Design & Accutronics).

7.6 Experimental Prototype

Experimental Prototypes allow developers to investigate, optimise and evaluate the mechanical properties of a product (Otto and Wood 2001). They are used to ascertain the feasibility of a product's working parts during the development stages of design. However, unlike Pre-Production Prototypes (Section 7.5), they do not resemble the final product and are used to obtain feedback on the functional performance (Ulrich and Eppinger 2003). While high in levels of fidelity and low in ambiguity on functional dimensions, they do not attempt to represent form or aesthetic aspects of the design concept. *Experimental Prototypes* are often low-cost and created quickly. The Experimental Prototype helps that helps define the layout or shape of a product, usually to replicate the actual product's physics. They are also known as Design-of-Experiment Prototypes (Figures 7.15–7.17).

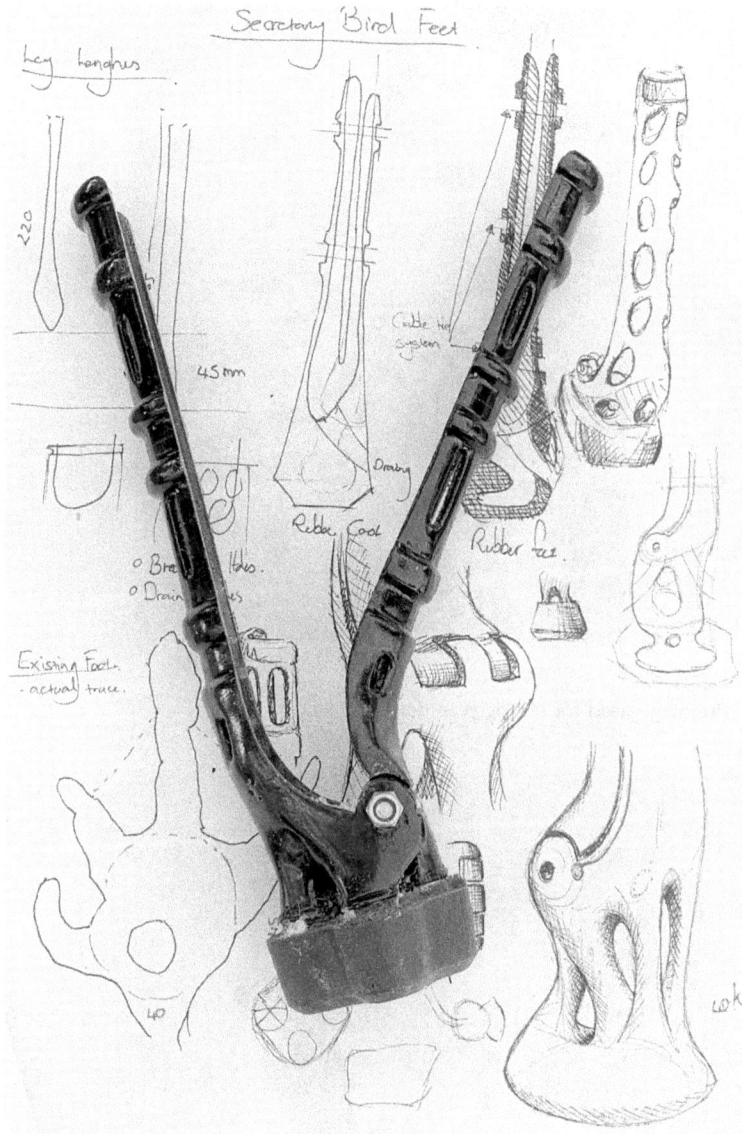

FIGURE 7.15
An Experimental Prototype with Reference Sketches (van der Walt, n.d.).

FIGURE 7.16
An Experimental Prototype used for testing (van der Walt, n.d.).

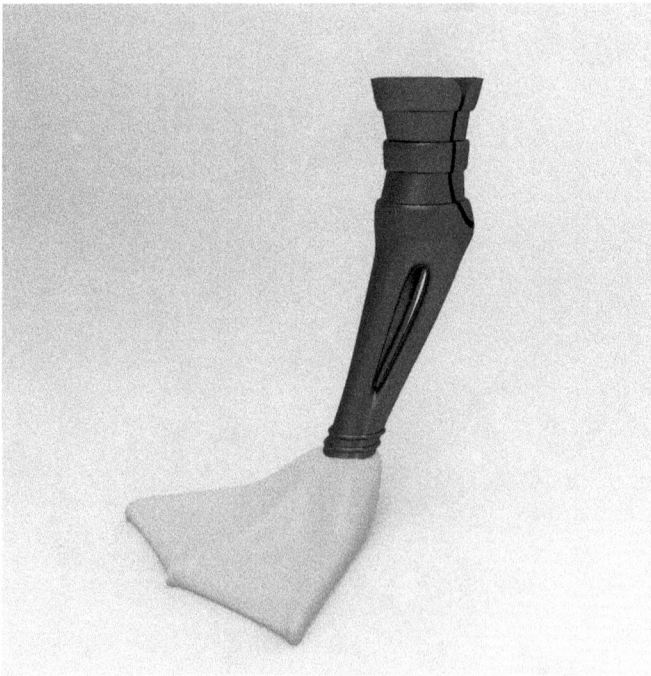

FIGURE 7.17
A Render of an Experimental Prototype (van der Walt, n.d.).

7.7 System Prototype

The *System Prototype* brings together the various working components of the product (Evans 1992). It integrates the parts as a system, allowing engineering designers to achieve a holistic functional representation of the design that can be tested according to its abilities (Otto and Wood 2001). They are used to test and validate technical functionality and to also provide opportunities for refinement and/or further optimisation. Similar to Experimental Prototypes (Section 7.6), *System Prototypes* allow interrogation of aspects and design features related to technical functionality, rather than aesthetic and/or use considerations.

In a *System Prototype*, off-the-shelf components may be used and the parts are roughly assembled. The*System Prototype* is a 3D design representation that combines the numerous components specified for the final product to test and assess functional aspects such as mechanism and performance (Figures 7.18 and 7.19).

FIGURE 7.18
A System Prototype (Choudhury, n.d.).

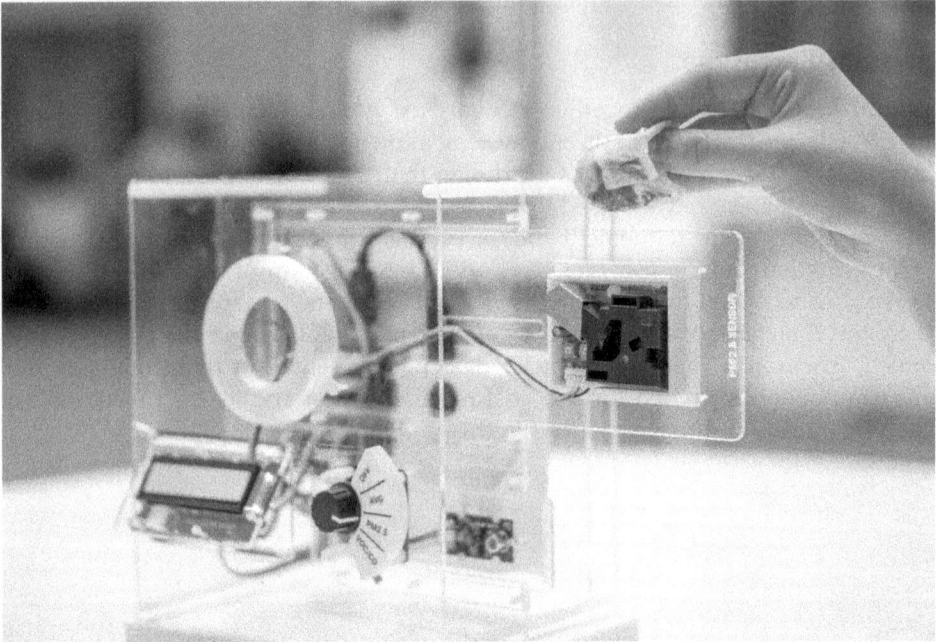

FIGURE 7.19
A System Prototype in use (Choudhury, n.d.).

7.8 Final Hardware Prototype

The *Final Hardware Prototype* is an integrated representation containing the final working parts as a whole and allows engineering designers and other stakeholders to discuss fabrication and assembly issues (Otto and Wood 2001). At this stage, the internal components are set in place without an exterior shell. They are different from Beta (Section 7.4) or Pre-Production Prototypes (Section 7.5) as a Final Hardware Prototype does not represent the exterior outlook and aesthetics of the design. Final Hardware Prototypes are thus specifically focused upon proving the viability of functionality in relation to hardware. In this sense, they represent the final design outcome unambiguously and completely prior to production . Therefore, they are often used at later phases of design development, as functional detail is refined and design developed towards manufacture and production. *Final Hardware Prototypes* are 3D visual design representations used to assist in the design and evaluation of product fabrication and other assembly issues (Figure 7.20, 7.21, and 7.22).

FIGURE 7.20
Fully functional Final Hardware Prototype incorporates prototype PCB & Electronics supplied by the client to be incorporated into the prototype casing (Hill, 2011; Cambridge Industrial Design for Osram).

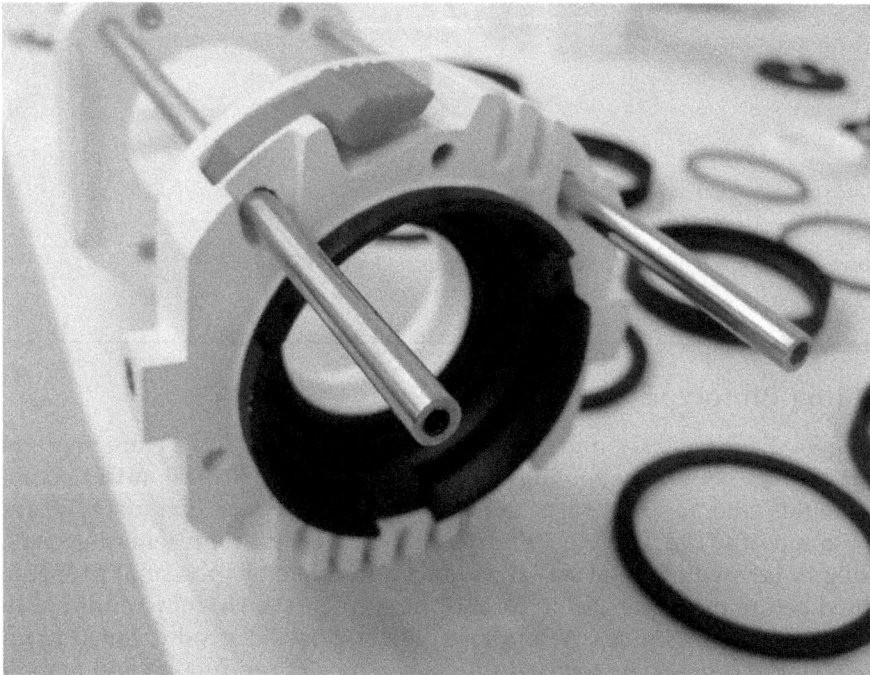

FIGURE 7.21
Final hardware prototype (Hill, 2011; Cambridge Industrial Design for Osram).

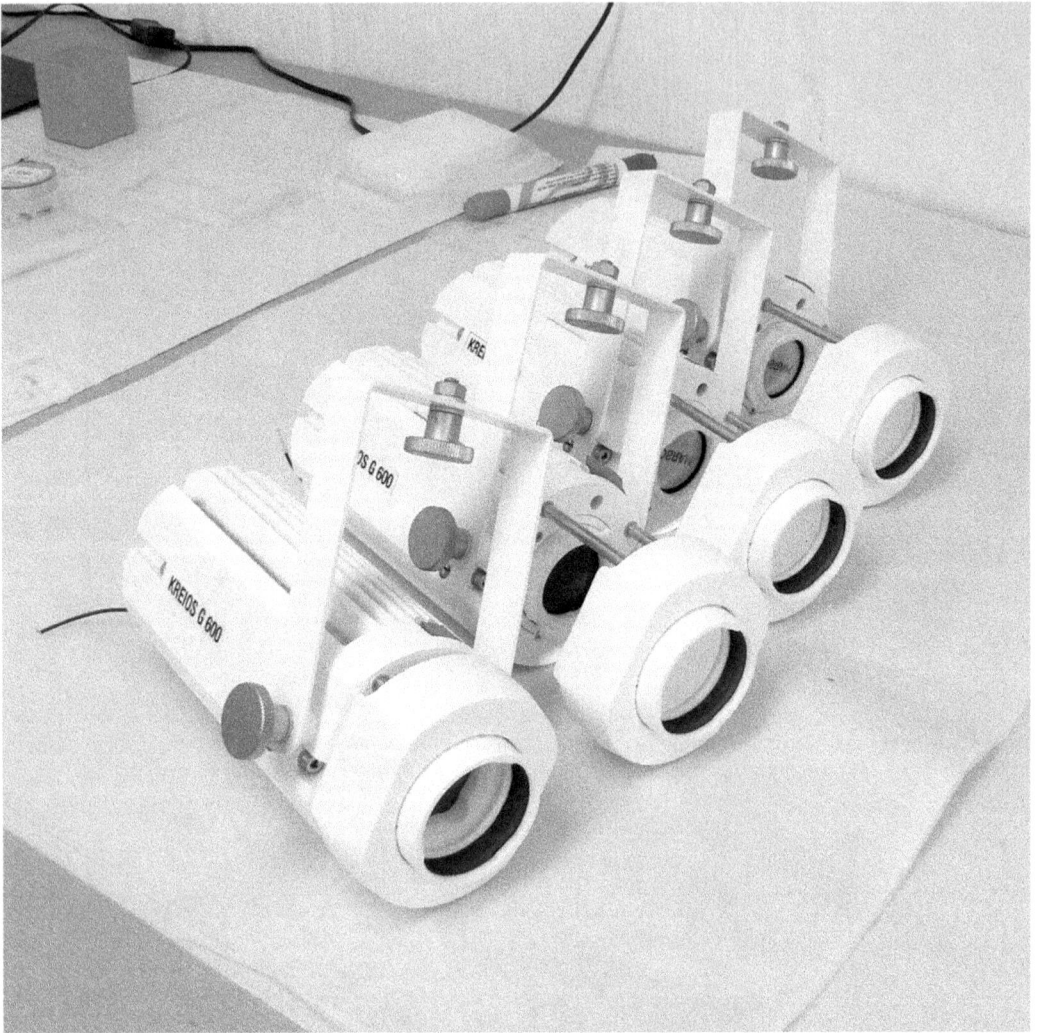

FIGURE 7.22
Final hardware prototype (Hill, 2011; Cambridge Industrial Design for Osram).

7.9 Tooling Prototype

According to Evans (1992), the *Tooling Prototype* is used to ensure that the steel pressings or die castings for tooling are correctly made. This minimises errors as incorrect moulds and tooling parts are hugely expensive to manufacture and very complex to modify. The Tooling Prototype is a 3D design representation that allows the tooling to be made for the actual product and to enable potential problems to be intercepted before discrepancies in form or fit occur. As such, they represent the final design outcome unambiguously and completely prior to production (Figure 7.23). *Tooling Prototypes* can also be focused upon processes of production and manufacture, rather than the design solution itself.

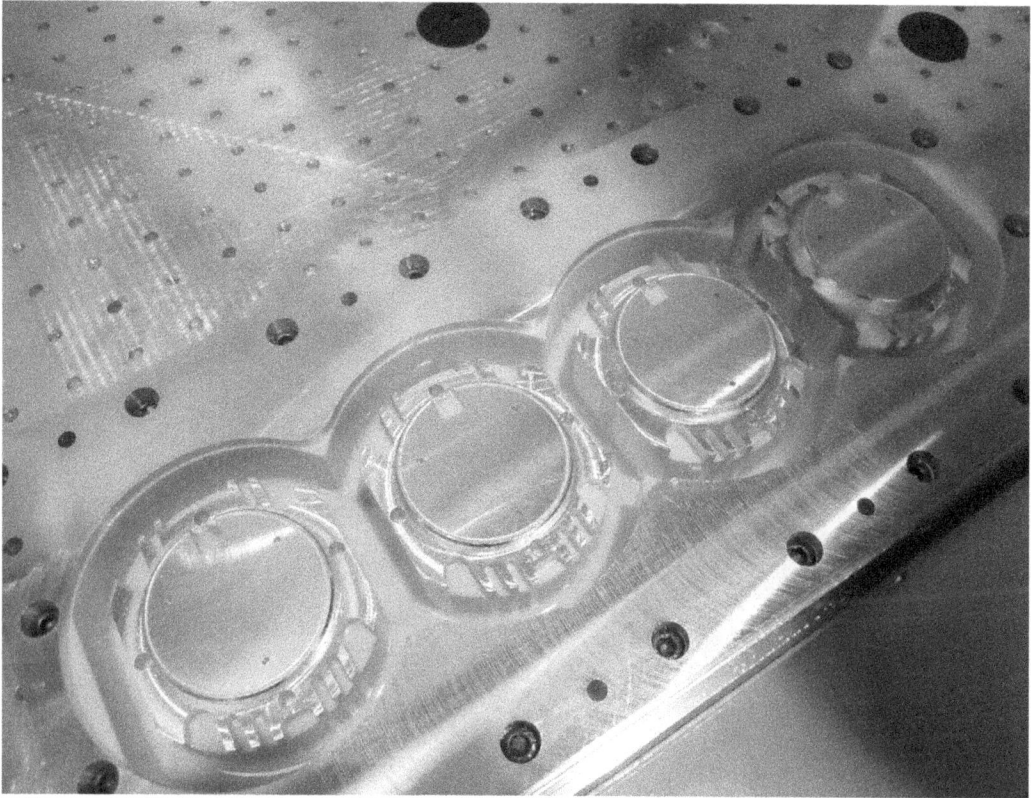

FIGURE 7.23
Prototype heatsinks, CNC machined, held in PU resin to prevent and resonance during machining (Hill, 2011; Cambridge Industrial Design for Osram).

7.10 Off-Tool Prototype

The *Off-Tool Prototype* consists of parts produced from the actual tooling and materials intended for the final product. They are mainly used by production engineers and engineering designers to validate component fit and assembly, while they may sometimes be used by industrial designers to check the finishing of parts (Evans 1992). Similar to a Tooling Prototype (Section 7.9), they are primarily focused on the testing and validation of production processes. In this, they aim to prove the viability of product components before the ramp-up of production proper. *Off-Tool Prototypes* are 3D visual design representations that consist of physical components produced from the actual tooling and materials intended for the final product (Figures 7.24, 7.25 and 7.26). These images show P-Series builds of the UNCells Chairman Mobile Phone using the first 1000 sets of production parts from each tool, such as injection moulded single and twin shot PCABS and Nylon 6 components. As per their use in practice, they represent design at the highest level of fidelity and are used in confirmation of design intent during final specification.

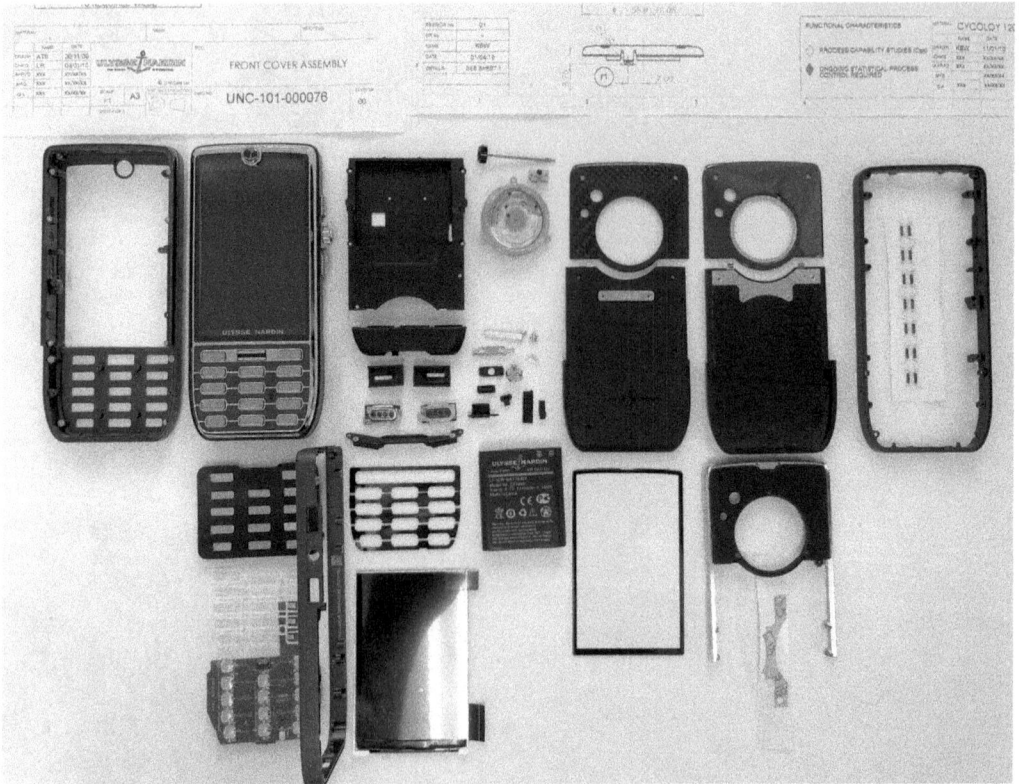

FIGURE 7.24
An Off-Tool Prototype UNCells Chairman P. Series mobile phone (Hill, 2010).

FIGURE 7.25
Part of Off-Tool Prototype of UNCells Chairman P.Series mobile phone (Hill, 2010).

FIGURE 7.26
Off-Tool Prototypes being produced for UNCells Chairman P.Series mobile phone (Hill, 2010).

7.11 Summary

This chapter outlines a fourth type of design representation (see also Chapters 4–6), discussing the role and use of *Prototypes* as representations of design intent (Section 7.1). We have discussed how Prototypes of various levels of fidelity are used through the process of design to support communication of design intent at various dimensions of interest. From works-like Prototypes to prototypical representations expressive of form and design aesthetic, the physical qualities of Prototypes have been described as providing increased sensorial experience towards potential design solutions. Unlike Models as design representations (Chapter 6), Prototypes offer increased width for exploring various functional and aesthetic dimensions of potential design solutions. This may be seen in the range of different Prototypes used across stages of design development and implementation through production.

The chapter has presented nine types of Prototypes often used in support of design practice. The Prototypes presented differ in terms of the focus of representation and level of fidelity in the expression of intent. Broadly the Prototypes presented and discussed may be classified as focused upon design form, appearance and aesthetic (Section 7.2), Prototypes used to integrate and validate functionality (Sections 7.6–7.8), Prototypes targeted at evaluating design in terms of process of manufacture (Sections 7.5, 7.9 and 7.10), and more holistic Prototypes that seek to represent intent on various functional and aesthetic dimensions (Sections 7.3 and 7.4).

Similar to the use of *Models* (Chapter 6), the classification of Prototypes illustrate the various ways they are used in practice to support the expression of intent to satisfy different communication, interrogation and design validation needs. In particular, the physical representation of intent is employed in practice as means to develop and validate various aspects of a potential design solution. Unlike *Models*, Prototypes are extensively used in support of design implementation through manufacture and production. Their ability to approximate various material and functional qualities at a high level of fidelity, allow for detailed validation of the production process before manufacture (Figures 7.27–7.29). In doing so, Prototypes move beyond other forms of design representation (i.e. Sketches, Drawings and Models) to communicate various functional and aesthetic aspects of a proposed design. However, although Prototypes are often in use during Detail Design, they are variously employed in support of Concept Design (i.e. *Appearance Prototype*) and Concept Development (*Alpha Prototype, Experimental Prototype*). The versatility of the design Prototype, in its ability to express and communicate various aspects of a design at different dimensions of interest (i.e. functional and/or appearance aspects), sees its use throughout the design process in support of development. This is also due to its ability to offer opportunities for communication through physical embodiments of intent. This physicality then provides clearer communication of various aspects in order to evaluate design potential at various stages of design development.

FIGURE 7.27
Prototyping in design process (Prototypum, n.d.).

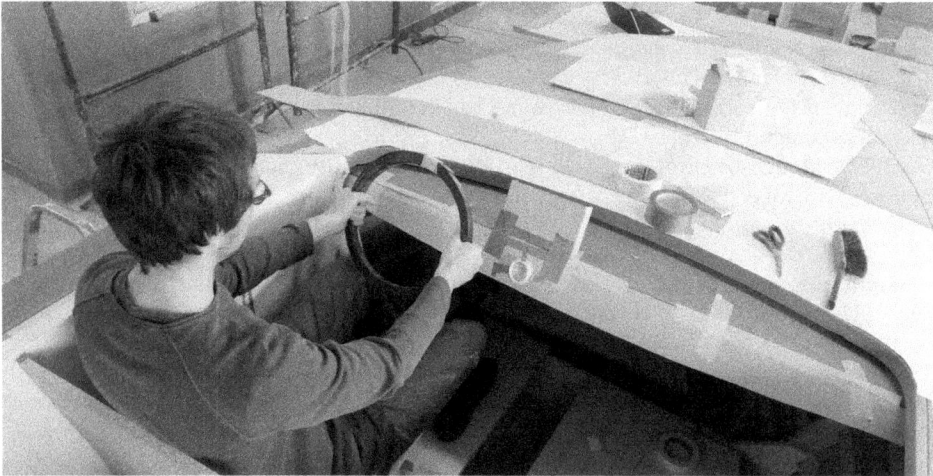

FIGURE 7.28
Using prototypes for the design process (Prototypum, n.d.).

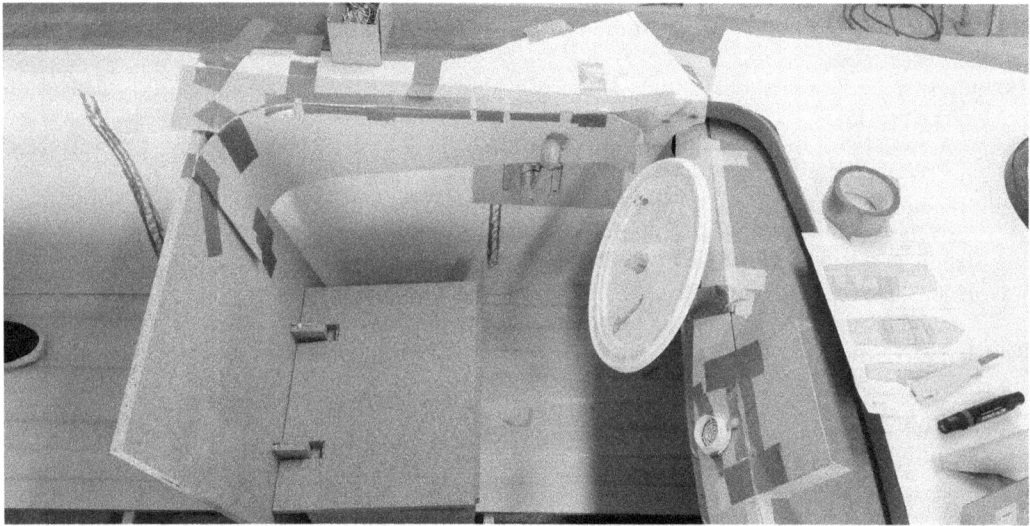

FIGURE 7.29
Using Drawings as a reference during the prototyping process (Prototypum, n.d.).

References

Avrahami, D. and Hudson, S. E. (2002) Forming Interactivity: A Tool for Rapid Prototyping of Physical Interactive Products. *Proceedings of the Conference on Designing Interactive Systems: Processes, Practices, Methods, and Techniques (DIS'O2)*, pp. 141–146.

Best, K. (2006) *Design Management - Managing Design Strategy, Process and Implementation.* Switzerland: AVA Publishing SA.

Bødker, S. and Buur, J. (2002) The Design Collaboratorium - A Place for Usability Design. *ACM Transactions on Computer- Human Interaction*, Vol. 9(2), pp. 152–169.

Budde, R., Kautz, K., et al. (1992) *Prototyping: An Approach to Evolutionary System Development*. Berlin: Springer. Quoted in: Yang, M. C. and Daniel, J. (2005) A Study of Prototypes, Design Activity, and Design Outcome. *Design Studies*, Vol. 26(6).

Evans, M. (1992) Model or Prototype Which, When and Why? *IDATER 1992 Conference*, Loughborough University (Design and Technology).

Evans, M. (2002) *The Integration of Rapid Prototyping within Industrial Design Practice*. Staff Thesis. Loughborough: Department of Design and Technology, Loughborough University.

Evans, M. A. and Campbell, R. I. (2003) A Comparative Evaluation of Industrial Design Models Produced Using Rapid Prototyping and Workshop-based Fabrication Techniques. *Rapid Prototyping Journal*, Vol. 9(5), pp. 344–351.

Finn, A. (1985) A Theory of the Consumer Evaluation Process for New Product Concepts. *Research in Consumer Behaviour*, Vol. 1, pp. 35–65. Quoted in: Söderman, M. (2002) Comparing Desktop Virtual Reality with Handmade Sketches and Real Products: Exploring Key Aspects for End-users' Understanding of Proposed Products. *Journal of Design Research*, Vol. 2(1).

Hanington, B. M. (2006) Interface in Form: Paper and Product Prototyping for Feedback and Fun. *Interactions* (January-February, Vol. 13(1), pp. 28–30.

Holbrook, M. B. and Moore, W. L. (1981) Feature Interactions in Consumer Judgements of Verbal Versus Pictorial Presentations. *Journal of Consumer Research*, 8 (1), pp. 103–113. Quoted in: Söderman, M. (2002) Comparing Desktop Virtual Reality with Handmade Sketches and Real Products: Exploring Key Aspects for End-users' Understanding of Proposed Products. *Journal of Design Research*, Vol. 2(1), pp. 7–26.

Holmquist, L. E. (2005) Prototyping: Generating Ideas or Cargo Cult Designs? *Interactions*, Vol. 12(2), pp. 48–54.

Kelley, T. (2001) Prototyping Is the Shorthand of Innovation. *Design Management Journal*, Vol. 12(3), pp. 35–42.

Kelly, T. and Littman, J. (2001) *The Art of Innovation - Lessons in Creativity from IDEO, America's Leading Design Firm*. London: Harper Collins Business.

Knoblaugh, R. R. (1958) *Modelmaking for Industrial Design*. New York: McGraw-Hill. In: Evans, M. A. (2002) *The Integration of Rapid Prototyping within Industrial Design Practice*. Staff Thesis. Loughborough: Department of Design and Technology, Loughborough University.

Kolodner, J. L. and Wills, L. M. (1996) Powers of observation in creative design. *Design Studies*, Vol. 17(4), pp. 385–416.

Kurvinen, E., Koskinen, I., et al. (2008) Prototyping Social Interaction, *Design Issues*, Vol. 24(3), pp. 46–57.

Luzadder, W. J. (1975) *Innovative Design*. London: Prentice-Hall. In: Evans, M. A. (2002) *The Integration of Rapid Prototyping within Industrial Design Practice*. Staff Thesis. Loughborough: Department of Design and Technology, Loughborough University.

Morrison, J. (1990) *Jasper Morrison - Designs, Projects and Drawings 1981–1989*. London: Architecture Design and Technology Press.

Otto, K. and Wood, K. (2001) *Product Design - Techniques in Reverse Engineering and New Product Development*. New Jersey: Prentice Hall.

Preece, J., Rogers, Y., et al. (2002) *Interaction Design - Beyond Human-computer Interaction*. New York: John Wiley and Sons.

Rosenberg, D. (2006) Revisiting Tangible Speculation: 20 Years of UI Prototyping. *Interactions*, January-February, Vol. 13(1), pp. 31–32.

Söderman, M.   (2002) Comparing Desktop Virtual Reality with Handmade Sketches and Real Products. Exploring Key Aspects for End-Users' Understanding of Proposed Product. *Journal of Design Research*, Vol. 2(1), pp. 7–26.

Sommerville, I. (1995) *Software Engineering, Wokingham*. England: Addison-Wesley. Quoted in: Yang, M. C. and Daniel, J. (2005) A Study of Prototypes, Design Activity, and Design Outcome. *Design Studies*, Vol. 26(6).

Subrahmanian, E., Monarch, I., et al. (2003) Boundary Objects and Prototypes at the Interfaces of Engineering Design. *Computer Supported Cooperative Work*, Vol. 12, pp. 185–203.

Ulrich, K. T. and Eppinger, S. D. (2003) *Product Design and Development* (3rd ed.). New York: McGraw-Hill.

Veveris, M. (1994) The Importance of the Use of Physical Engineering Models in Design. *IDATER 1994 Conference*, Loughborough University (Design and Technology).

Virzi, R. A., Sokolov, J. L., et al. (1996) Usability Problem Identification Using Both Low and High Fidelity Prototypes. *Proceedings of CHI '96*, pp. 236–243.

Ward, A., Liker, J. K., Sobek, D. and Cristiano, J. (1995) The Second Toyota Paradox: How Delaying Decisions Can Make Better Cars Faster. *Sloan Management Review*, Vol. 36(3), pp. 43–61.

Yang, M. C. (2009) Observations on Concept Generation and Sketching in Engineering Design. *Research in Engineering Design*, Vol. 129(5), pp. 1–11.

Yang, M. C. and Daniel, J. (2005) A Study of Prototypes, Design Activity, and Design Outcome. *Design Studies*, Vol. 26, pp. 649–669.

8

Case studies and conclusions

8.1 GMC Case Study 01

In January 2008, the UK arm of the 'Global Machinery Company' or GMC commissioned Solve 3D to produce a Reference Model of their concept for the next-generation hydrogen fuel cell power tools. The Model was of a cordless drill and charging station. The *Appearance Model* was for a power tool exhibition in Cologne and had to demonstrate all of the materials that would be used in the final manufacturing process, including a representative hydrogen fuel cell cartridge (Figures 8.1 and 8.2). In production, the unit was to be a combination of anodised aluminium pressure die

FIGURE 8.1
Appearance Model of Hydrogen Charger & Drill, illuminated to demonstrate the User interface (Hill, 2008 for GMC).

DOI: 10.1201/9781003227694-8

FIGURE 8.2
Another view of an Appearance Model (Hill, 2008 for GMC).

castings, brushed stainless steel and gloss acrylic. Part of the brief was to bring the Model to life, with a fully illuminated control array but the power stipulation was for an integral rechargeable battery supply so that the unit would function continuously for over 8 hours.

Working with lead designer John Hogarth, Solve 3D decided that this would have to be a project spread across its three sister companies, Solve 3D, Rapid 3D and 3D Definitive Coatings (3DDC). Parts were built from the process of using Vat Photopolymerisation (also known as Stereolithography Apparatus or SLA) with 'Accura Blue Stone' for rigidity, then post cured to withstand a surface temperature up to 200 degrees for post-processing operations. CNC machined, polyurethane parts were electroform-plated, first with a 120 µm copper layer and then with a layer of technical nickel plating to give a 'cold' feel. A glass bead blasted finish was then applied to represent the surface finish of a pressure die casting. A hand brushed finish was applied to represent the areas of stainless steel (Figures 8.3–8.5). The process took 54 hours from start to finish, approximately 2-3 weeks faster than quoted lead times to produce parts using a traditional lost-wax foundry process.

The brief for the main charger frame was to express the look and cold feel of brushed stainless steel and contain the deep embossed logo of the company. The mainframe was CNC machined and fabricated from high-density epoxy tooling board. Then using a series of chemical pre-treatments, 150 µm of copper followed by silver nickel was

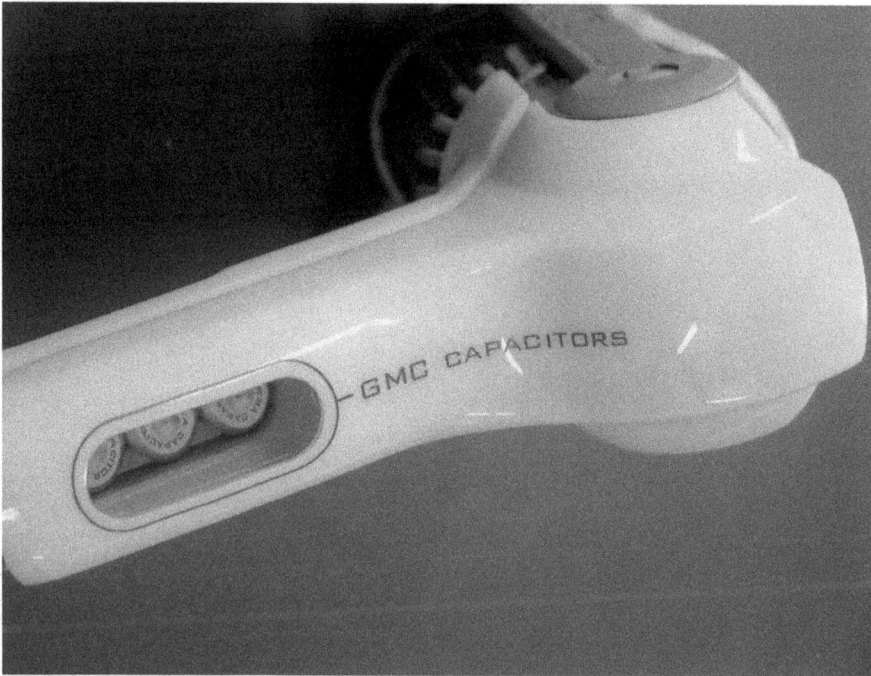

FIGURE 8.3
Parts of an Appearance Model, a combination of plated PU & CNC machine and polished acrylic (Hill, 2008 for GMC).

FIGURE 8.4
Finished Components for Appearance Model (Hill, 2008 for GMC).

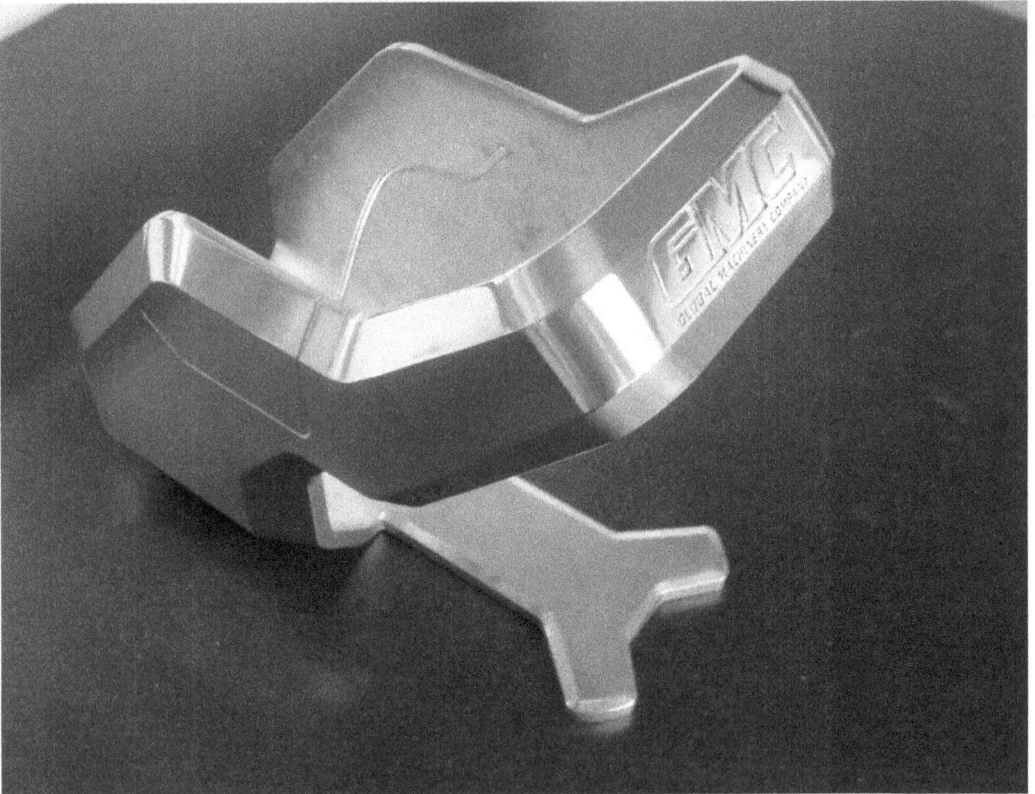

FIGURE 8.5
Copper base finish applied prior to nickel plating for Appearance Model (Hill, 2008 for GMC).

applied to the parts. This was then hand brushed followed by vinyl graphics being applied. This process was also used for a number of drill components with a final plated finish of 'Technical Black' applied to the drill chuck and the hydrogen-gel cartridge. The plastic components were Computer Numerical Control (CNC) machined from solid billets of cast white acrylic, hand polished and graphics applied. The hydrogen gel was represented as tinted, uncured liquid silicone which was encapsulated in an acrylic flask Figure 8.6 Figure 8.8. The final electronic lighting was a circuit made up of low voltage LEDs linked to a rechargeable lithium mobile phone battery. The total build time for the *Appearance Model* was 3 weeks and was a unique and successful blend of using cutting-edge Additive Manufacturing and traditional model making (Figures 8.6, 8.7, and 8.8).

FIGURE 8.6
Range of Material finishes for appearance model. (Hill, 2008 for GMC).

FIGURE 8.7
Fully assembled Appearance Model (Hill, 2008 for GMC).

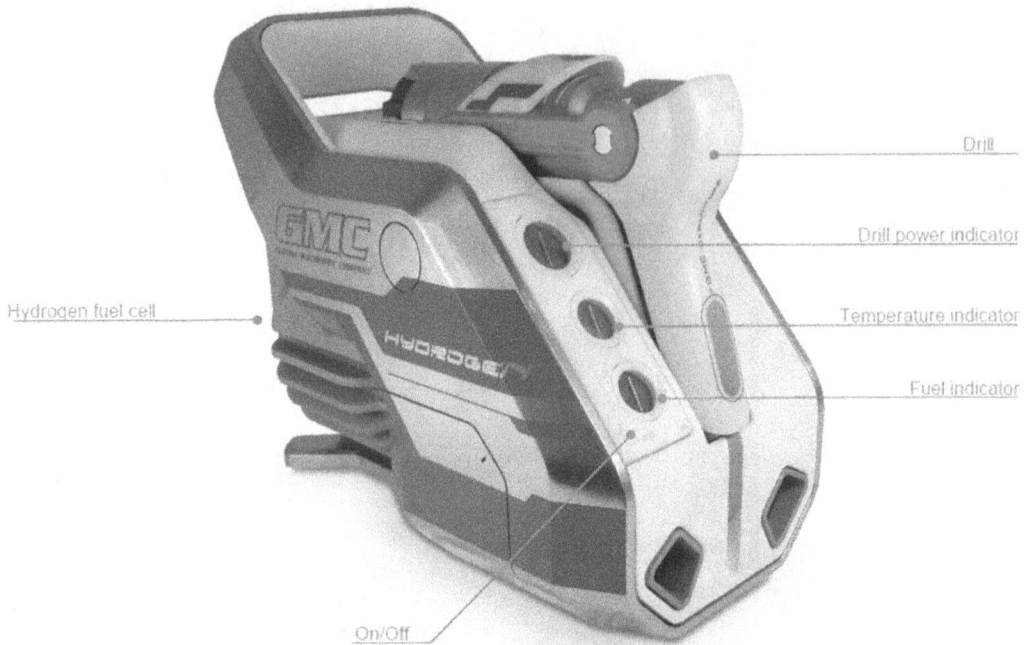

FIGURE 8.8
Presentation Drawing using a photograph of an Appearance Model (Hill, 2008 for GMC).

8.2 Mojavi Case Study 02

Mojavi was a final year project designed by Thomas Mortimer at Brunel University London in 2017. It was developed as a home hub unit, utilising low-cost Light Detection and Ranging (LiDAR) electronics as a means of enabling hand gestures to control home devices. This project had an emphasis on working electronics and Tom analysed pre-built circuit boards and LiDAR sensors to identify which were suitable for his project. By building multiple *Design Development Models*, he was able to resolve the technical constraints, including potential working distances, environmental lighting conditions, user movement and product placement, resulting in the use of the VL53L0X ranging sensor to achieve a functional *System Prototype*. During the process, it was important to find the correct distance and optimum angle at which the user would interact with the product. This required a *Concept of Operation Model* to be produced during the process of development, consisting of a built-in laser pointer to identify the maximum angle that the device would be able to function. An *Appearance Prototype* was produced, based on Arduino Uno, a Time-of-Flight (TOF) sensor, 2D gesture sensor and light sensor that resulted in further fine tuning of the refresh rate and accuracy of the system.

After proving that the electronics worked, Tom went on to contact a Printed Circuit Board (PCB) manufacturer as the chip needed a selection of extra components to function. This resulted in a custom-made electronic circuit board that reduced the footprint of the product, making the enclosure for the *Final Hardware Prototype* more compact and easier to assemble with integrated electronics (Figures 8.9–8.12).

FIGURE 8.9
Concept of Operation Model (Mortimer, 2017).

FIGURE 8.10
Circuit board for the Concept of Operation Model (Mortimer, 2017).

FIGURE 8.11
3D CAD model of the enclosure (Mortimer, 2017).

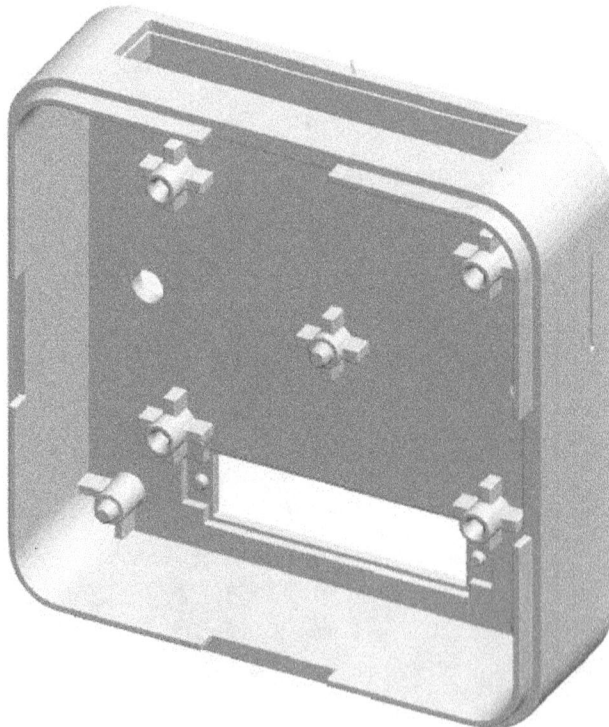

FIGURE 8.12
Updated 3D CAD model of the enclosure (Mortimer 2017).

With a good understanding of the electronic requirements, Tom investigated the form factor of the product, in which out of the three initial *Concept Drawings*, the cube was determined to be the most suitable. The design of the enclosure was created in Solidworks that considered aspects of design for manufacturing, with the use of injection moulding where separate parts would then be produced. The multiple iterations of the CAD design were subjected to Finite Element Analysis (FEA) to ensure that the thickness of the enclosure was suitable (Figures 8.13–8.23). The final design of the enclosure was produced using Additive Manufacturing to simulate injection moulding and to create an accurate *Appearance Model* as well as to produce parts for the *Final Hardware Prototype*.

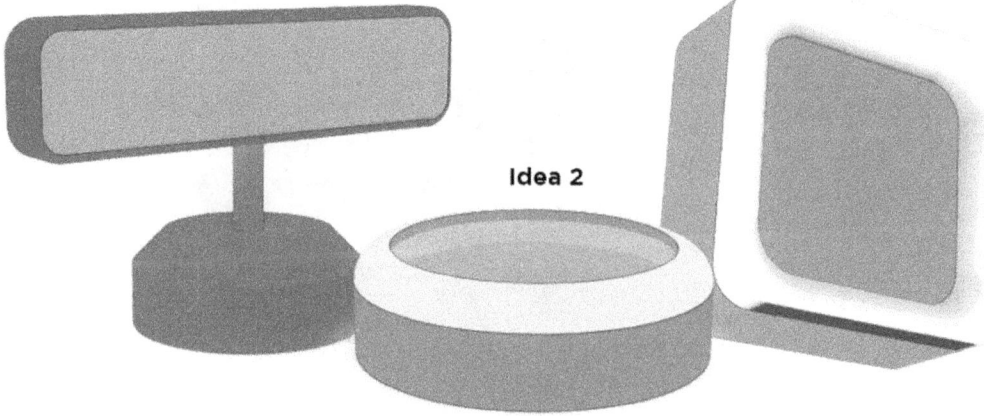

FIGURE 8.13
Iterations of the design produced using 3D CAD (Mortimer, 2017).

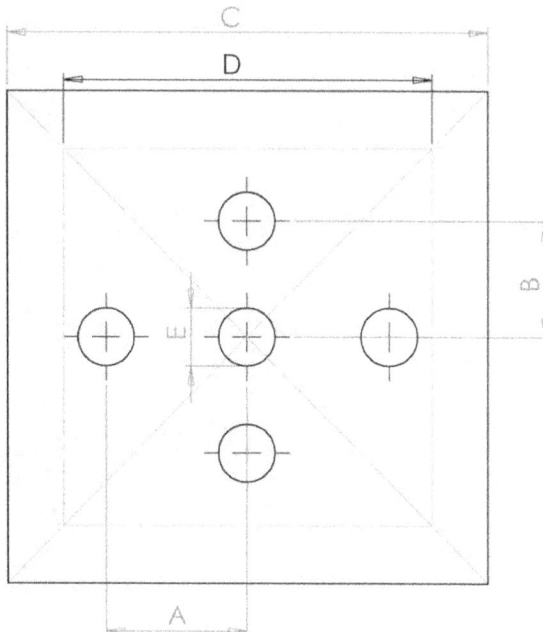

FIGURE 8.14
A Single-View Drawing (Mortimer, 2017).

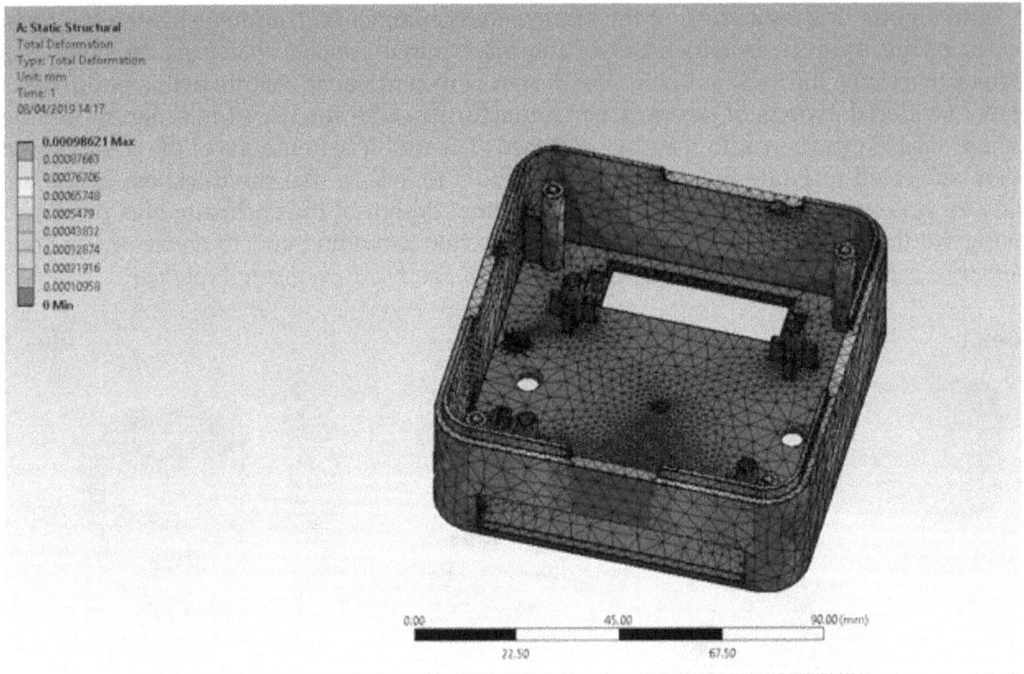

FIGURE 8.15
Finite Element Analysis using 3D CAD (Mortimer, 2017).

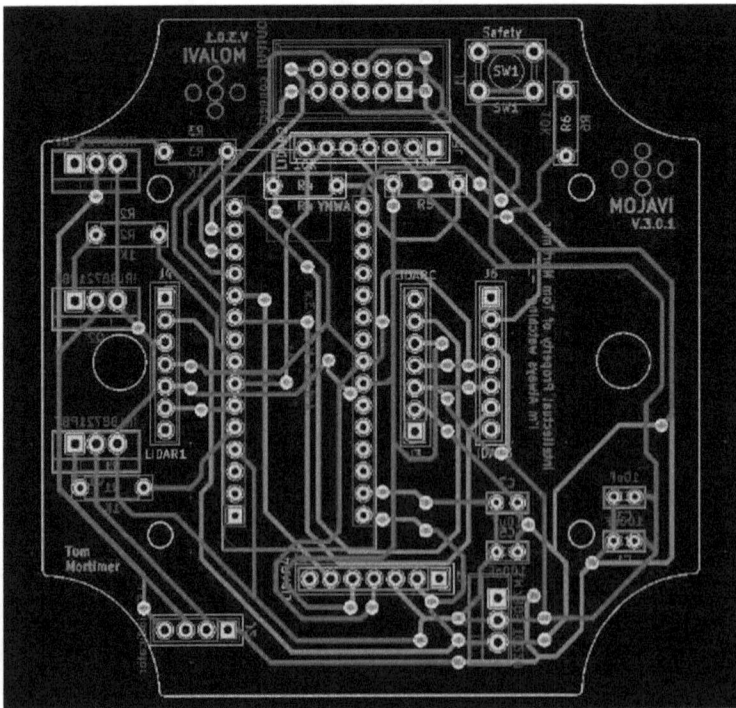

FIGURE 8.16
Design of a circuit board (Mortimer, 2017).

FIGURE 8.17
Circuit board being produced (Mortimer, 2017).

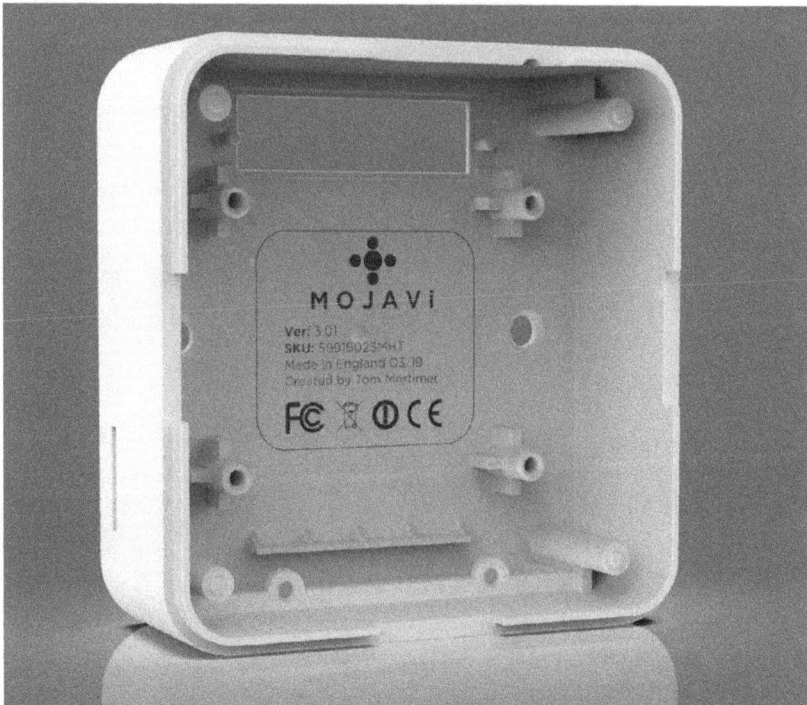

FIGURE 8.18
Rendering of the enclosure (Mortimer, 2017).

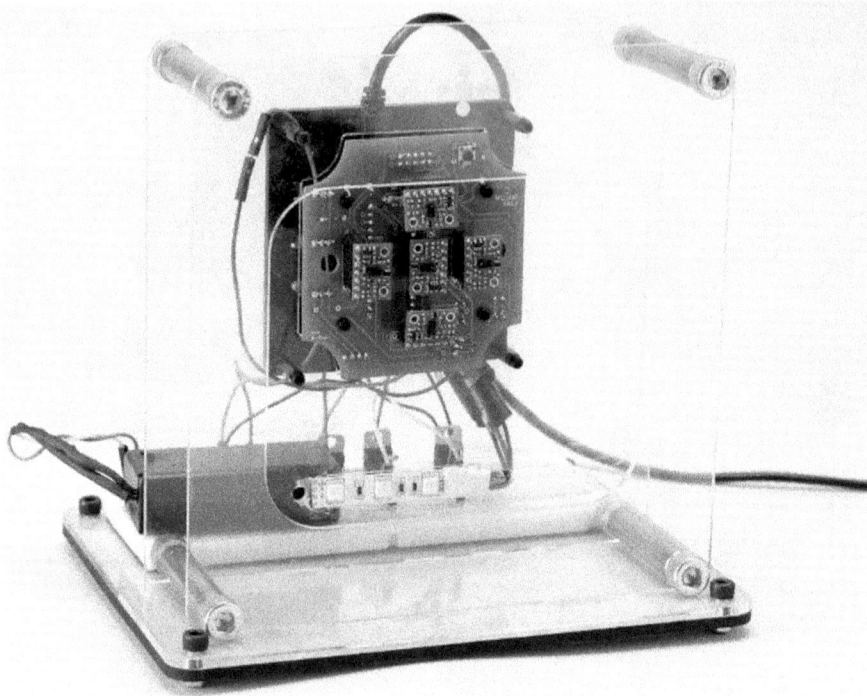

FIGURE 8.19
System Prototype (Mortimer, 2017).

FIGURE 8.20
Parts of the Production Concept Model (Mortimer, 2017).

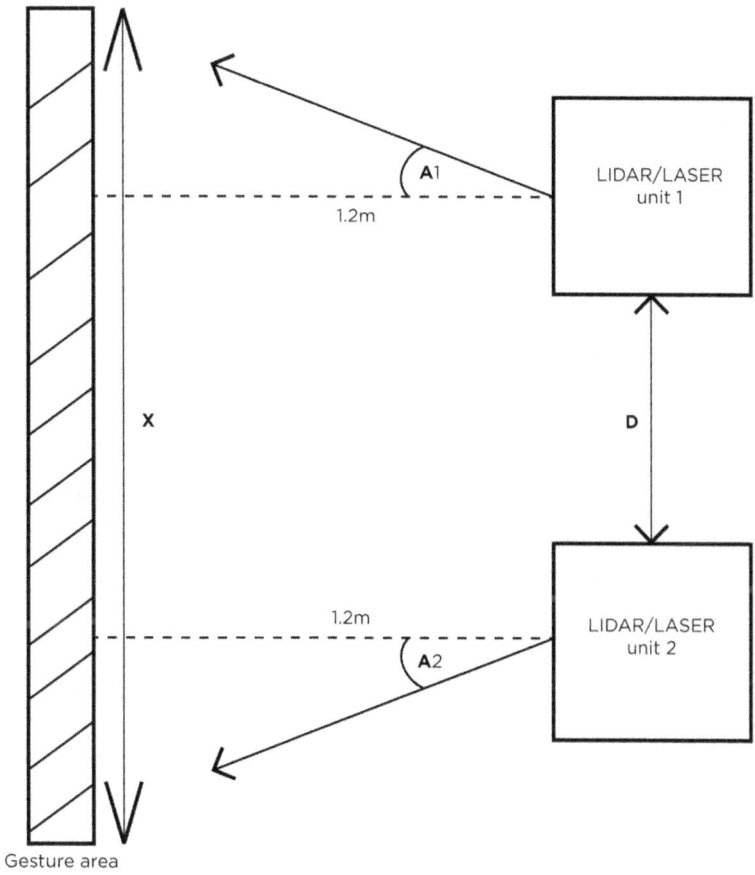

FIGURE 8.21
A Diagrammatic Drawing (Mortimer, 2017).

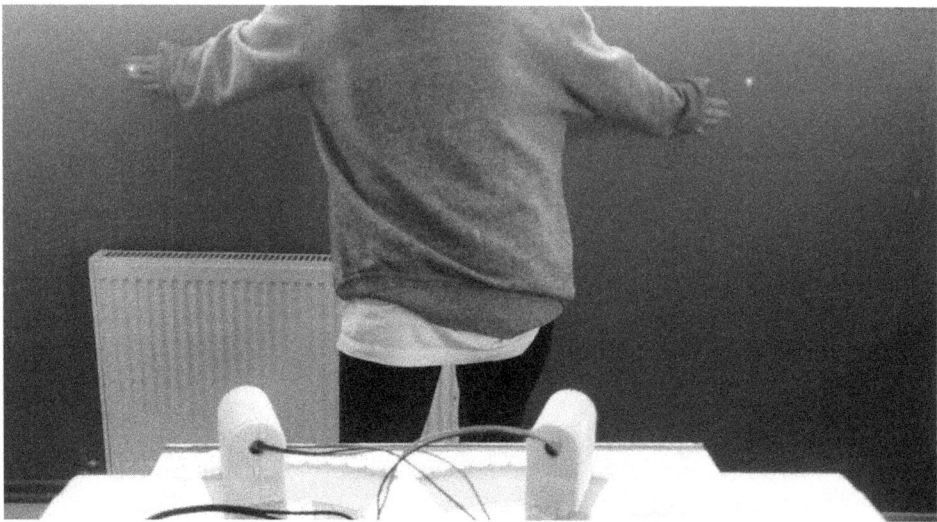

FIGURE 8.22
Tests using a Functional Concept Model (Mortimer, 2017).

FIGURE 8.23
Appearance Prototype of the product (Mortimer, 2017).

8.3 Deeptime Case Study 03

The Spirula Speakers was designed in 2019 by Deeptime, a design and tech studio based in Prague, Czech Republic. It was initially conceived in 2015, but Additive Manufacturing technology at that time was expensive and did not offer process reliability, quality or the expected sound quality. The first *Appearance Model* was produced using Woodfill filament (Figure 8.24) and it took the firm a few years of experimentation with *Production Concept Models* to achieve the process of hardening the semi-finished printed product into the silica sand composite to ensure the desired material rigidity, durability and perfect acoustic properties (Figures 8.25 and 8.26). The legs were manufactured using different materials such as aluminium and brass from milling and other surface treatments (Figure 8.27).

The use of Binder Jetting as a process of Additive Manufacturing with sand allowed the company to produce nature-inspired shells printed from a single piece and due to its anti-resonant and absorption properties. As a result of using Additive Manufacturing, the firm was able to produce multiple *Pre-Production Prototypes* for acoustics testing, fine tune the tolerances for assembly, as well as to optimise the number of electronic components and parts for the final product (Figure 8.28–8.30).

FIGURE 8.24
Appearance Model (Deeptime, 2019).

FIGURE 8.25
A Production Concept Model (Deeptime, 2019).

FIGURE 8.26
Production Concept Models (Deeptime, 2019).

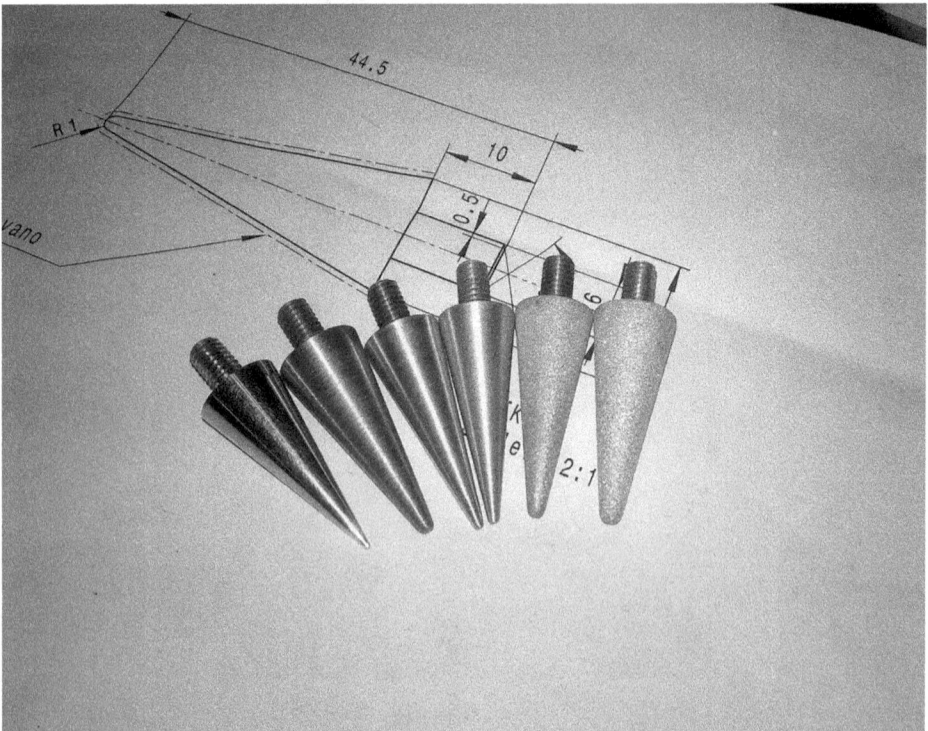

FIGURE 8.27
Parts for the Production Concept Model (Deeptime, 2019).

FIGURE 8.28
Components for a Pre-production Prototype (Deeptime, 2019).

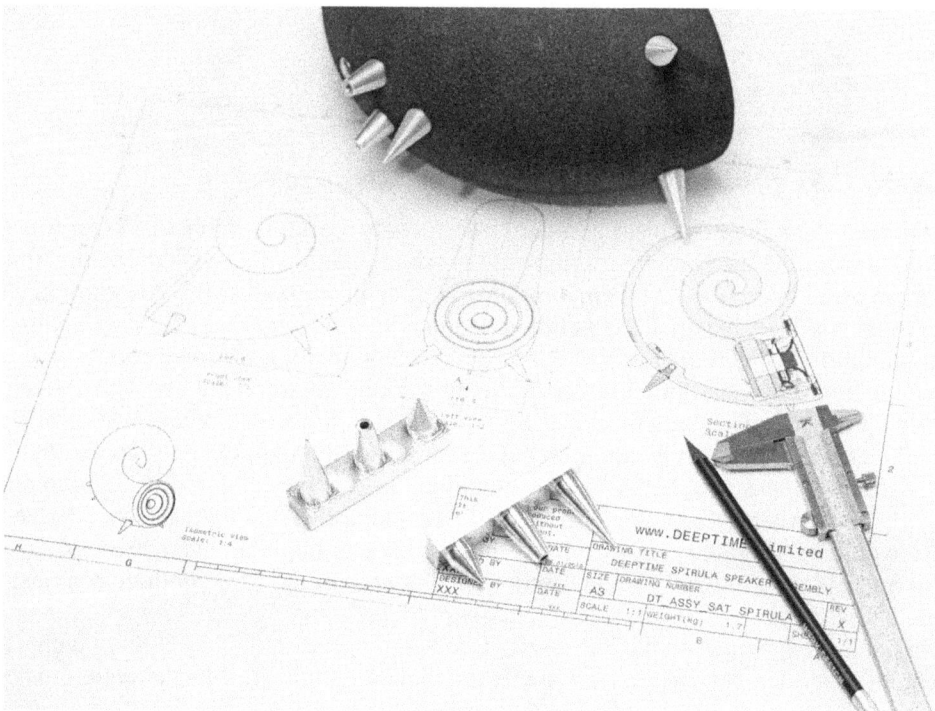

FIGURE 8.29
Technical Drawing and Pre-Production Prototype parts (Deeptime, 2019).

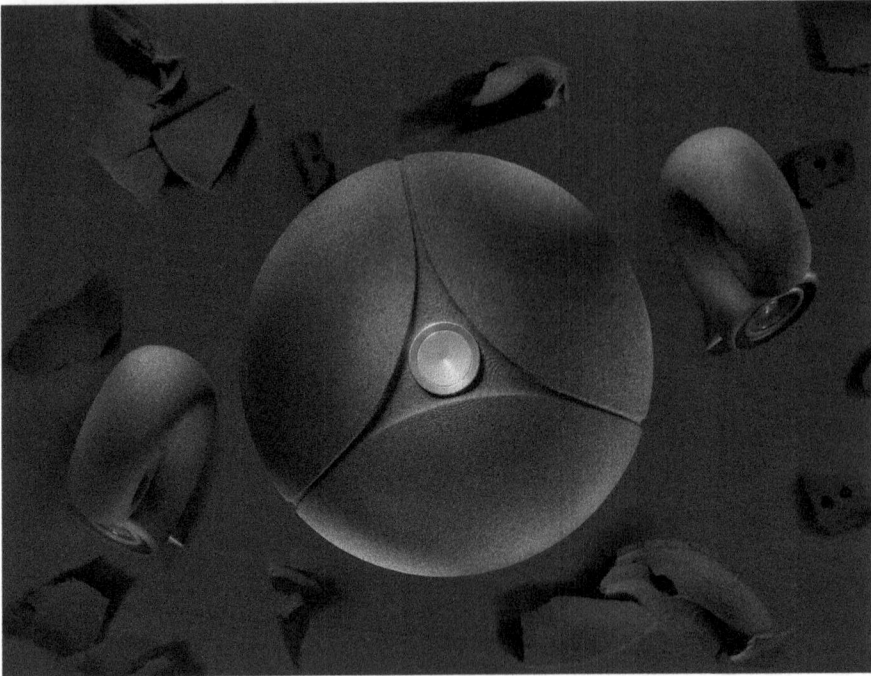

FIGURE 8.30
Pre-production prototype components (Deeptime, 2019).

8.4 Aero Case Study 04

For his final year major project at Brunel University London, Ajharul Choudhury developed a portable air quality monitoring product. The device would feature in-built sensors to provide accurate, real-time reading of specific air pollutants. After undertaking research, it was decided that the product would monitor PM2.5, PM10 and volatile organic compounds (VOCs). Research into potential sensors and components was conducted to find the most suitable components for the intended function. An electronic *Functional Concept Model* was created using Arduino due to its ease of use and coding simplicity. The major components, consisting of a dust sensor that measures the concentration of particulates in the air, and a VOC sensor would be used, in addition to an LED screen and LED light ring. Adafruit Neopixel was used to program the LED light where a lighted sequence would be activated every time a new air quality rating is received every 20 seconds. The overall *System Prototype* was encased in a Plexiglas frame as a product demonstrator (Figure 8.31–8.34).

After finalising the functional components, Choudhury produced *Study Sketches* with a view of defining the form for the product enclosure. Solidworks CAD Models were created to

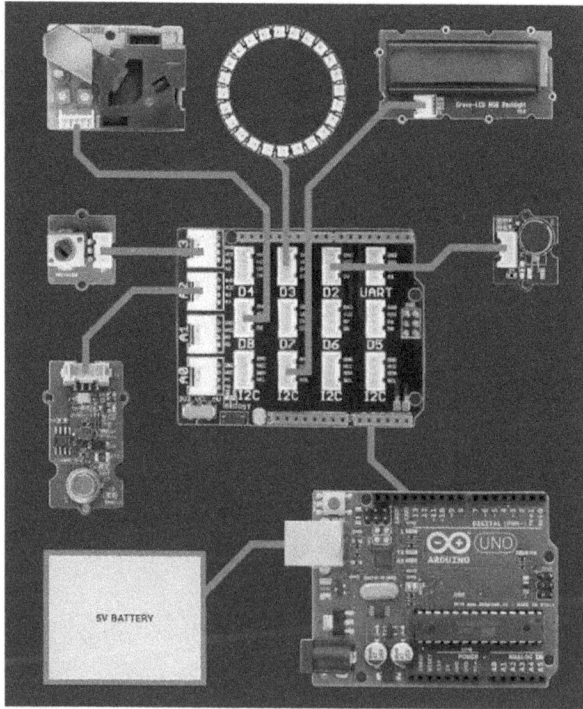

FIGURE 8.31
Systems prototyping process (Choudhury, n.d.).

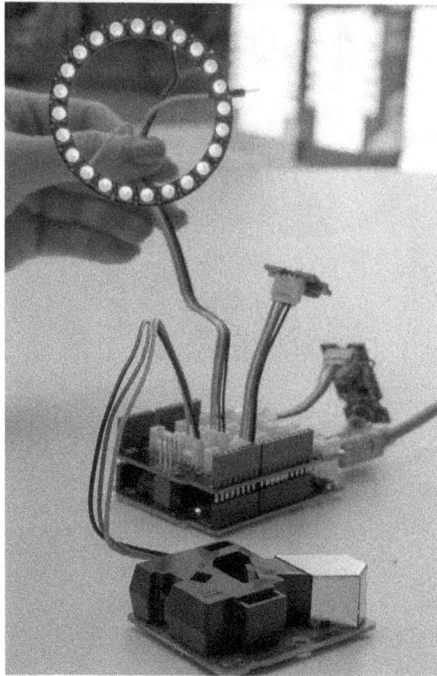

FIGURE 8.32
A Systems Prototype (right) (Choudhury, n.d.).

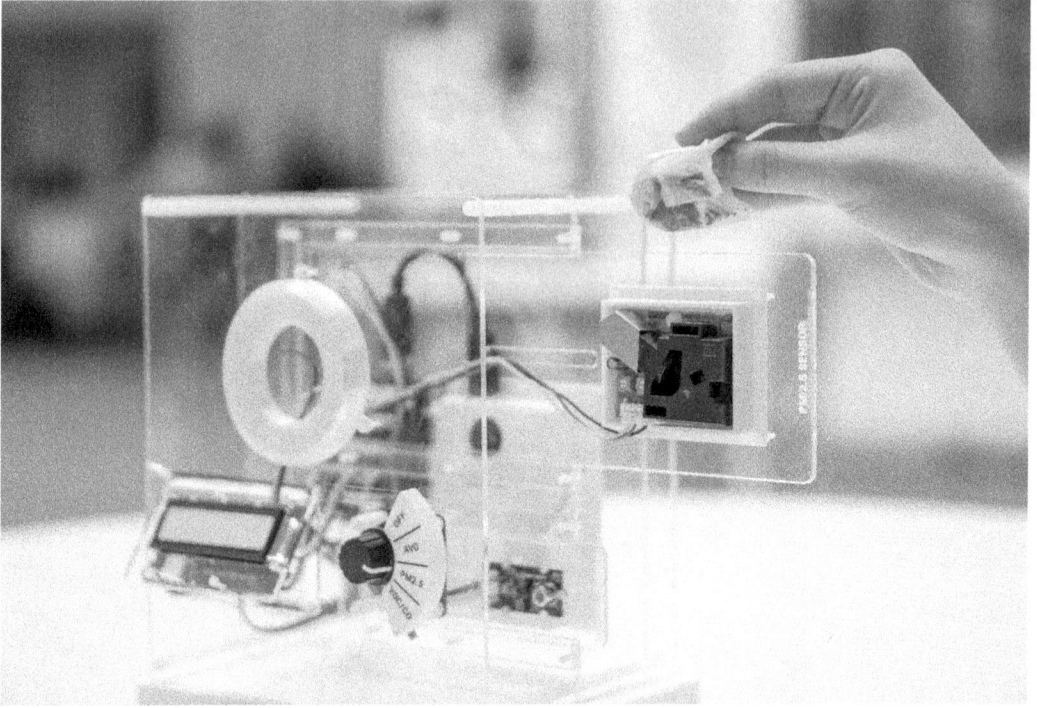

FIGURE 8.33
Systems Prototype in use (Choudhury, n.d.).

FIGURE 8.34
An Appearance Model (Choudhury, n.d.).

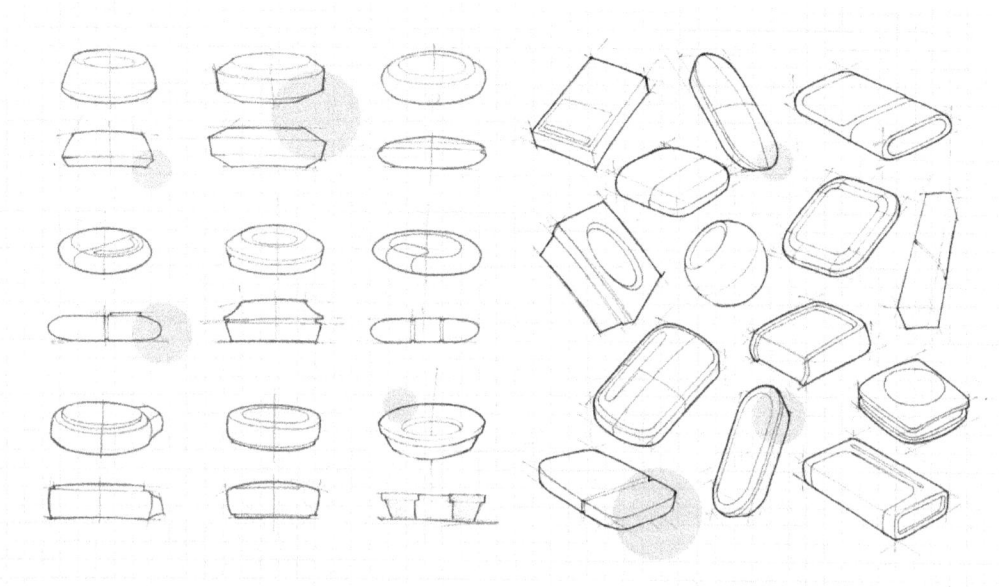

FIGURE 8.35
Study Sketches (Choudhury, n.d.).

represent a more accurate realisation in which the Models were fabricated using a combination of additively manufactured parts and manually produced components. The 3D CAD Models were subject to FEA, ensuring that the wall thicknesses and internal supports would be sufficiently robust for handling the *Appearance Prototype* (Figure 8.34–8.41).

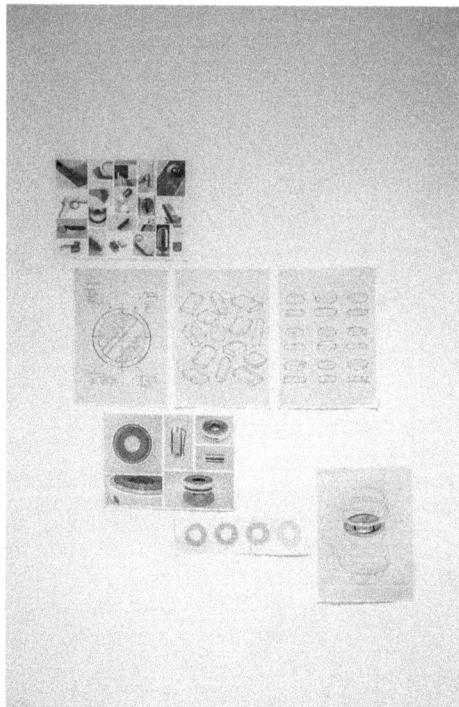

FIGURE 8.36
Conceptual design ideation using Study Sketches (Choudhury, n.d.).

COMPONENTS

· BATTERY
· LED LIGHTS
· LCD SCREEN
· VIBRATION
· SWITCH
· CHARGING PORT
· BLUETOOTH

POLLUTANT
SENSORS

· PM 2.5
· PM 10
· VOC
· CO
· NO₂

ATTACHMENTS

· CLIP FOR TRAVEL
· DESK CHARGE /HUB
· WALL CHARGE /HUB

OTHER SENSORS

· TEMPERATURE
· HUMIDITY
· LOCATION

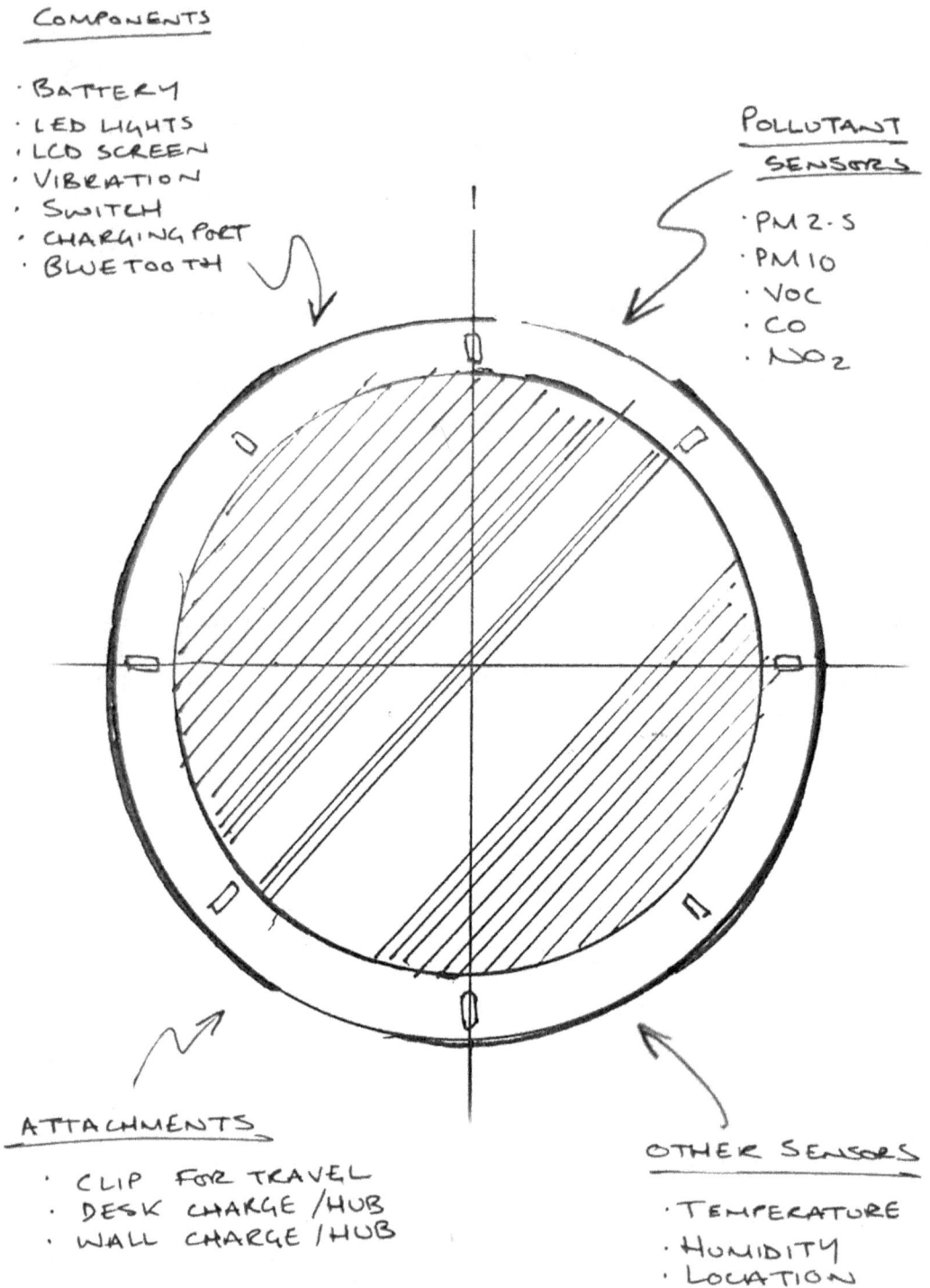

FIGURE 8.37
A Prescriptive Sketch (Choudhury, n.d.).

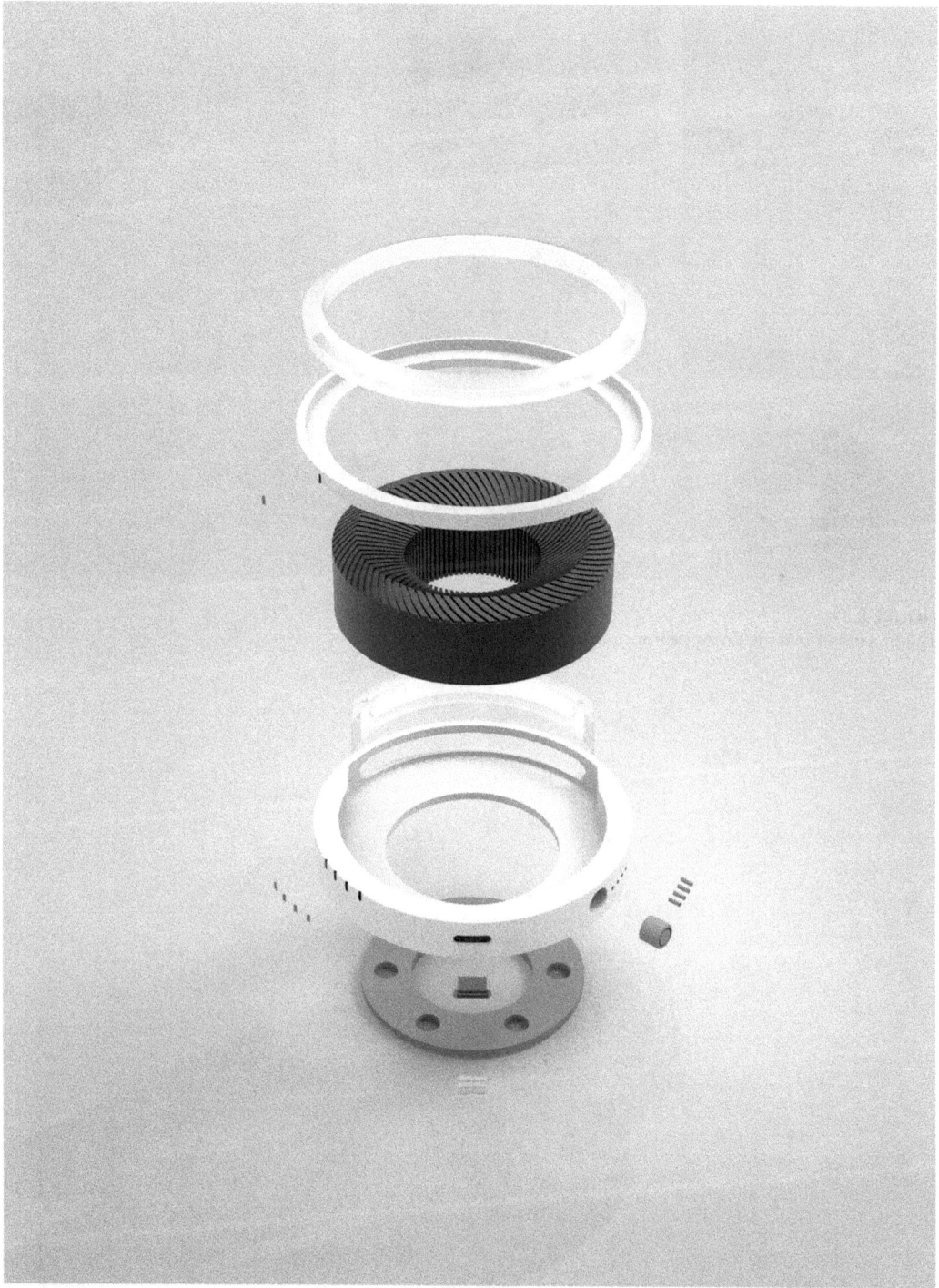

FIGURE 8.38
A Technical Illustration showing an exploded view of parts (Choudhury, n.d.).

FIGURE 8.39
The process of making an Appearance Model (Choudhury, n.d.).

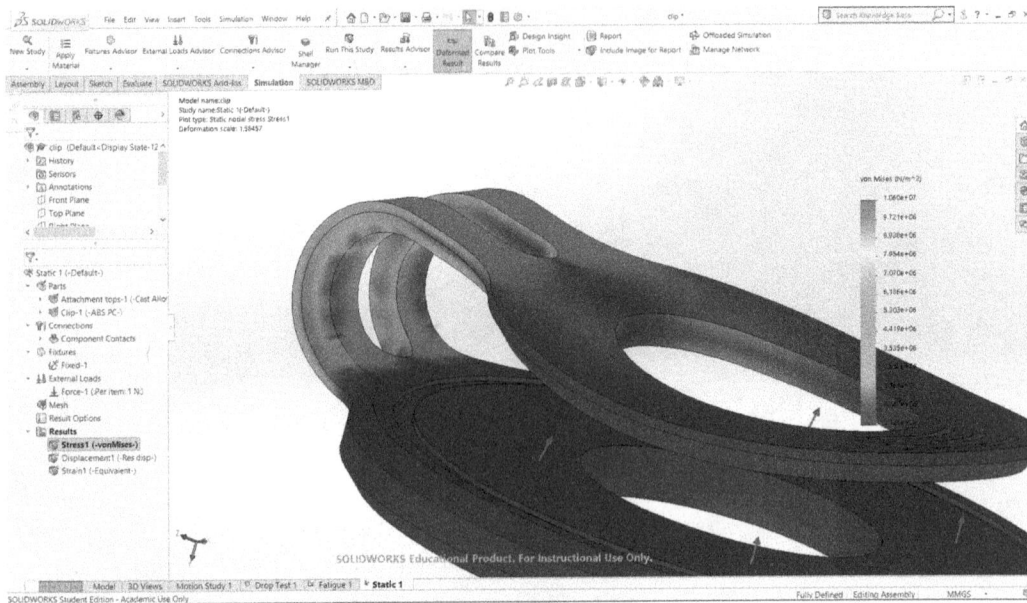

FIGURE 8.40
Using Finite Element Analysis in 3D CAD (Choudhury, n.d.).

FIGURE 8.41
Final Appearance Model (Choudhury, n.d.).

8.5 Conclusions

Throughout this book, we have made the case for our thesis that design representations are critical for the practice of design, and the unique approach of design thinking in practice. To achieve this, we have explored and discussed design representations in terms of its historic roots in industrial design, its application as part of a process of design, and design representations' relation to ill-defined design problems, their paired solutions and the types of thinking involved in an activity of design framed by the reflective-practice paradigm first proposed by Schön (1983).

In Chapter 1, we introduced ideas and concepts towards more clearly defining the practice of design. In this definition, we positioned design practice as an exploration of possibilities towards the identification of design solutions aimed at resolving often complex design problems, achieved through a process of problem-solution pairing. We then looked at how the use of design representation may locate within design practice, before turning towards more recent works aimed at defining a particular type of creative cognition, now popularly termed as *Design Thinking*. Within the context of design thinking, we positioned design representation as a critical component of what it means to engage design thinking during practice. The chapter further contextualised design representation through reference to both its historic roots in the industrial revolution, and its ubiquitous use to support design activity. Design representations' role in a process of design was also considered with reference to key stages of design that iteratively move from *Concept Design*, through *Concept Development* and into *Detail Design*.

Chapter 2 continued and expanded our discussion of design representations' role in and support for design thinking. The chapter introduced a discussion of how design representation is influenced by the media of expression (i.e. analog sketch representation, compared to CAD). In doing so, we explored the potential for an interaction effect between the media of expression and design representation. The tool of design representation will implicate the representations made, and design thinking during practice. However, this interaction may often be reduced or increased through the designer's expertise in tool use, or lack thereof. Expanding our discussion of design thinking and representation, subsequent subsections (Section 2.2) examined design representation through the lens of embodied cognition. Embodied/distributed cognition was considered in terms of its relationship to the reflective-practice paradigm for understanding cognition during design practice (Schön 1986). From here, the construct representational fidelity (Section 2.2.3) and representational ambiguity (Section 2.2.2) were operationalised as possible candidates to define the various types of design representation used in practice, their different influence on design thinking, design communication, and their role in relation to a design process described as moving, iteratively, from *Concept Design*, through *Concept Development* and into *Detail Design*.

Chapter 2 concluded with a discussion of relations between design representation, the nature of ill-defined design problems and the search for appropriate problem-solution pairs. Foregrounded by a comprehensive treatment of relations between design thinking, design representation and design practice, the chapter ended with a general definition of design representation based on three related aspects (Section 2.5). First, design representation was defined as the expression of design intent towards potential design solution candidates. Second, we restricted our definition of design representation to the expression of intent that attempts to approximate the potential design on one or more dimensions of interest through iconic approximation of a final design deliverable (i.e. looks/works like representations). Although we do not deny the importance of other modes of design representation (i.e. through text, the spoken word, gesture, etc.), this book is focused on representations that attempt to express the physical, functional and/or semantic features of potential design solutions. They attempt to approximate the properties of propositional design solutions, and their development. Third, design representation was discussed as a type of creative distributed cognition between design thinking and design practice. This representation is seen to provide unique opportunities to first identify, then explore and finally develop potential solution candidates in response to ill-defined design problems. The first section of our definition provides a holistic context as a scaffold for the definition. That is, design representations are expressions of intent towards solution ideas. The second aspect aimed at framing the scope of our definition of design representation as physical and/or functional approximations of solution ideas, in contrast to other forms of expression (i.e. verbal, written, gestural). The third positions design representation as a type of distributed cognition, supporting design thinking as ill-defined design problems are engaged, and solution candidates identified, explored, developed and finally refined.

As in Chapter 1, Chapter 2 aimed to support our position that design representation is critical to design thinking. Further, the chapters described the particular types of cognition involved in the use of design representation, and how considering the nature of design representation through theoretical constructs operationalised in our definition (i.e. *ambiguity, fidelity*) may help understand relationships between design thinking, design practice and the broader process of design within which representations are employed and applied (i.e. *Concept Design, Concept Development, Detail Design*). In Chapter 3, we turned to the applied practice and activity of design to present various types of design representation (Section 3.2). Here, design representation was discussed in terms of its relation to the purpose of design (Section 3.1), and the tools and media used in the construction of design representations (Section 3.3). In particular, we focused upon 2D and 3D tools of representation, as well as digital and manual media of expression.

A discussion in Chapter 3 on types of design representation was further extended in four subsequent chapters, each describing a particular type of representation . Within each of these four chapters, sub-types of design representations were presented and discussed in relation to design thinking, and communication of solution ideas. These four chapters provide an extensive and holistic taxonomy of design representations used in support of design practice. Each chapter included sub-types of representation described within each of the four chapters such as a Presentation Drawing (Figure 8.42).

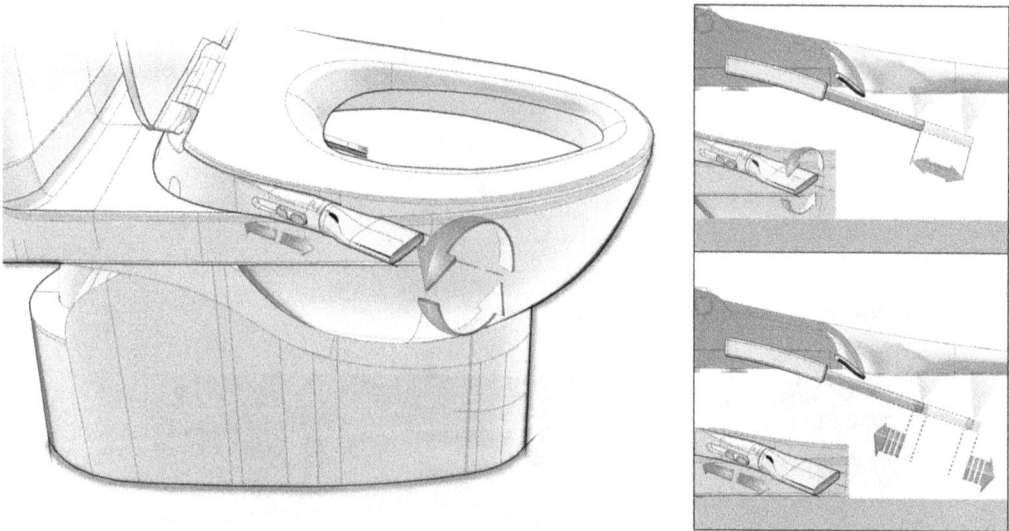

FIGURE 8.42
A Presentation Drawing explain the use of a product (Chan, n.d.).

To ground our taxonomic classification of design representations (Chapters 4–7), we presented case studies describing the use and application of representations in support of design practice. In each case, (GMC Case Study 01, Mojavi Case Study 02, Deeptime Case Study 03 and Aero Case Study 04), the application, role and importance of representation types are highlighted through discussion of their support for design conceptualisation, development and implementation during a process of design. This discussion included their role and use in identification of possible solutions and role as distributors of cognition in design thinking, as well as their support for design practice.

In this book, we have attempted to address our argument for design representation as critical to design practice. This was achieved through the discussion of design representation in terms of design thinking, design process and the various types of representation often used to support design practice. We have discussed design representations and the design process, and how design representations support both solution identification and development, as well as communication between stakeholders. In particular, we have discussed how

FIGURE 8.43
Photo-realistic Rendering of a product (Choudhury, n.d.).

FIGURE 8.44
Close-up view of an Appearance Model (Choudhury, n.d.).

distributed cognition, together with ambiguity/fidelity in representation, may provide an opening towards understanding relationships between design cognition and representation of design intent. This may then contribute to the development of generalisable theory of design thinking, as critically related to design representation. In terms of the design process, we have explored and discussed representation's relation to a staged model of process, wherein the types of representations used relate to and are a reflection of the evolving requirements of design development.

The changing character of design representations, from ambiguity to high-fidelity in the expression of intent, is further evidence of a critical relationship between the representation, design thinking and practice. Throughout the current work, we have attempted to describe design representation's role in design practice from various perspectives. In order to continue to build general theory of design that underpins design as practice, as cognition and as process, more research is now required. This work has helped us to develop a foundational understanding towards the critical relationship between design representation, design thinking and practice. How might a distributed approach to understanding design thinking contribute to theory on design? What is the nature of any relationship between process and representation in different contexts, when pursuing various design problems and in contrast to other forms of cognition? How might a more holistic understanding of design representation support work in adjacent fields of study, creative cognition, innovation, decision-making sciences and the cognitive sciences

More work is required to further develop an understanding of an interdependent relationship between design representation, thinking and practice. Understanding this relationship has the potential to build theory useful in application for design education and industry. Our book has provided points of departure for the continued work towards understanding the role and use of design representation.

Reference

Schon, D. (1983) *The Reflective Practitioner*. London: Ashgate.

Index

Italic type denotes figures.

A

activity and design representation, 4–7, *5–7*
additive manufacturing (3D printing), 19, *20*, 23, 167
Aero case study 04, 202, 206, *207–208*
aesthetic design, 15
alpha prototype, 174–175, *175*
ambiguity, xvii, 52, 55
American Society of Mechanical Engineers (ASME), 127
AMF, 78
appearance models, *11*, *18*, *79–80*, *92–93*, 149, 152–153, *154*, 174, 189, *190–193*, 202, *203*, *208*, 212–213, *216*; using additive manufacturing (3D printing), *80*
appearance prototypes, *16*, 173–174, *173–174*, 202
assembly concept models, 157–159, *159–160*

B

"back-talk", 4–5
Bauhaus school of design, 15
beta prototype, 176, *176*
boundary object, 66

C

case studies and conclusions, 189–217; GMC case study 01, 189, *190–193*; Mojavi case study 02, 194–202; Deeptime case study 03, 203, *203–206*; Aero case study 04, 202, 206, *207–208*
coded sketches, 55, *55*, 114–115, *115*
comprehensive prototypes, 172
Computer-Aided Design (CAD), 19, 32, 44, 45, 77–78, *78–79*, *94*, 96; formats to save data in, 78
Computer-Aided Industrial Design (CAID) xcii, 19, 94
Computer-Aided Manufacturing (CAM), 19
concept design, xvii, 24–30, 60
concept development, xvii, 30–31, *31–33*, 61–63
concept drawings, 129–130, *129–130*
concept of operation model, 155–156, *156*, 194, *194–195*
conceptual design ideation using study sketches, *209*
conclusions, 206, *209–213*, 210–211, 213–217, *215–216*
coordinate artifact, 66
cross-section view, *75*

D

Dareshani, Natasha, 42, *42*, *92–93*
Deeptime case study 03, 203, *203–206*
Dell computers, 15, *16*
Dermachi, Valentina, 43, *43*, *64*, *75*, *77*, 136
design: definition of, 1; as iterative process, *6*
design aesthetics, 3
design and design representation, 1–40; design as practice, design as an activity, 1–7, *2–3*; activity and design representation, 4–7, *5–7*; design thinking and design representation, 8–13; industrial design, 13–23; history of, 15–17; work of, 17–23; stages of new product development, 23–37; concept design, 24–30; concept development, 30–31, *31–33*; detail design, 31–37, *34–36*
design development model, 150–152, *152–153*, 194
design drawings, 127
design fidelity, xvii
design ideation meeting, *82*
design process, 8, *8–9*; models of, 27, *68*
design representation: definition of, xvii, 68, *69*
design representation in practice, 73–103; digital 2D media, 88–91; digital 3D media, 92–96; manual 2D media, 88, *89*; manual 3D media, 92, *92–93*; purpose of design representation, 73–80; summary, 96–97; tools of design representations, 84–88; types of design representations, 81–84
design semantics, 3
design talk-back, xvii
design thinking and design representation, 8–13
design thinking through representation, 41–72; design representation, 41–46; as construction, 45–46, *46–48*; and media of expression, 43–45, *45*; design

representation: a definition, 68, *69*; representation and design cognition, 46–58; design representation and ambiguity, 52–55, *53–55*; design representation and fidelity, 55–58, *57–59*; design representation as reflective-practice, 48–52, *49–51*; representation and design process, 58–66; design representation: concept design, 60, *61–62*; design representation: concept development, 61–64; design representation: detail design,
64–66; representations, design problems and solutions, 66–68; summary, 69–70
detail design, xvii, 31–37, *34–36*
diagrammatic drawings, 132–133, *134*, *201*
diagrams, 127
Diaz, Oscar, 58, *58–59*
digital: definition of, xvii, 89
digital 2D media, 88–91
digital 3D media, 92–96
digital media, xvii
digital rendering, *106*
drawings, 4, 127–146; *see also* specific types of drawings; concept drawings, 129–130, *129–130*; as design representation, 127–129, *128*; diagrammatic drawings, 132–133, *134*; drawings as design representation, 127–129, *128*; general arrangement drawings, 137, *138*; multi-view drawings, 135–137, *137*; presentation drawings, 130–131, *131*; scenarios and storyboards, 131–132, *132–133*; single-view drawings, 133–135, *135–136*; summary, 141–145, *143–144*; technical drawings, 137–140, *139–140*; technical illustrations, 140–141, *141–142*
dual processing model of design, 12, *12*

E

evolutionary prototypes, 170–171
experimental prototypes, *35*, 43, *43–44*, 178, *179–180*
explanatory sketches, 107
explorative sketches, 107

F

fidelity, 56–58

final element analysis, *198*
final hardware prototypes, 182, *182–183*
final prototypes, 177
finite element analysis in 3D CAD, *212*
first prototypes, 174
focused prototypes, 172
focus group, *28*
functional concept models, *3*, *33*, *75*, *86*, 148–149, 153–155, *155–156*, 201

G

general arrangement drawings, 137, *138*
Global Machinery Company (GMC) case study 01, 189, *190–193*
glossary of terms, xvii–xviii

H

Hayashi, Saeka, 55, *55*
high-fidelity prototypes, 172
Hogarth, John, 189
hydrogen fuel cell power tools, 189

I

idea sketches, *5–7*, *42*, *42*, *51*, 83, *83*, *89*, 108–109, *109–110*
ideation workshop, *21*
ill-defined design problems, xviii
ill-defined solutions, xviii
incremental prototypes, 170–171, *171*
industrial design, xviii, 13–23; history of, 15–17; work of, 17–23
Industrial Designers Society of America (IDSA), 16
information sketches, *17–18*, *54*, *57*, *63*, *108*, 115–116, *116–117*
inspiration sketches, 47, 119, *119–120*
intermediary objects, 66
investigative sketches, 107

J

Jameel, Azim, 52, *52–54*

L

Light Detection and Ranging (LiDAR) electronics, 194
Lim, Mark, 20
low-fidelity prototypes, 172

M

manual: definition of, xviii, 88
manual 2D media, 88, *89*
manual 3D media, 92, *92–93*
manual media, definition of, xviii
McClumpha, Matthew, 1, *2–3*
medium/media, definition of, xviii
memory sketches, 113–114, *114*
memo sketches, 109
mobile phone, *18*
mock-ups, *16, 18*
models, 4, *58, 69,* 147–165; *see also* specific types
    of models; appearance model, 152–153,
    154; assembly concept model, 157–159,
    159–160; concept of operation model,
    155–156, *156*; design development
    model, 150–152, *152–153*; as design
    representation, 147–150; 3D sketch
    model, 150, *151*; functional concept
    model, 153–155, *155–156*; iterative
    improvement of, *59*; production
    concept model, 157, *157–158*; service
    concept model, 159–162, *161–162*
Mojavi case study 02, 194–202
Mortimer, Thomas, *78–80, 134, 160, 162,* 194,
    194–202
multi-view drawings, *45, 56, 64, 77,* 135–137,
    137

N

napkin sketches, 109
new product development, xviii
Nugent, Spencer, 56, *57, 61, 63, 83*

O

off-tool prototypes, *34,* 184, *185–186,* 186
operation model, *36*

P

Paper Prototyping (Snyder), 88
"pencils before pixels", 81
persuasive sketches, 107
pilot-production prototypes, 177
pre-production prototypes, *34,* 177–178,
    177–178, 205–206
prescriptive sketches, 107, 119–121, *122–123, 210*
presentation drawings, *91,* 127, 130–131, *131, 215*
principle models, 155

production concept models, 157, *157–158, 200,*
    202, 204
product renderings, *107*
prototypes, 4, *15, 31, 58, 73,* 167–188; alpha
    prototypes, 174–175, *175*; appearance
    prototypes, 173–174, *173–174*; beta
    prototypes, 176, *176*; experimental
    prototypes, 178, *179–180*; final
    hardware prototypes, *84,* 182, *182–183*;
    off-tool prototypes, 184, *185–186,* 186;
    pre-production prototypes, 177–178,
    177–178; prototypical design
    representations, 167–172, *168–172*;
    summary, 186–187; system prototypes,
    178–179, *181*; tooling prototypes, 182,
    184, *184*
prototypical design representation, 167–172,
    168–172
purpose of design representation, 73–80

R

rapid prototyping, 167
reference models, *185–186*
referential drawings, 127
referential sketches, *46,* 111–113, *112–113*
reflection-in-action, xviii
reflective-practice, xviii
renderings, *17–18, 29, 32, 47–48, 63, 65–66, 76,*
    116–118, *118, 199*; photo-realistic, *215*
representation and design cognition, 46–58;
    design representation and ambiguity,
    52–55, *53–55*; design representation
    and fidelity, 55–58, *57–59*; design
    representation as reflective-practice,
    48–52, *49–51*
representation and design process, 58–66;
    design representation: concept design,
    60, *61–62*; design representation:
    concept development, 61–64; design
    representation: detail design, 64–66
representations, design problems and
    solutions, 66–68

S

Samsung Airpen, 15–16, *16*
scenarios and storyboards, *27,* 131–132, *132–133*
service concept models, 159–162, *161–162*
Shah, *69, 85–86, 113*
single-view drawings, *5, 75, 106,* 133–135,
    135–136, 197
sketches, 4, 6, 105–125; *see also* specific types of

sketches; coded sketch, 114–115, *115*; complexity in, 74; idea sketch, 108–109, *109–110*; information sketch, 115–116, *116–117*; inspiration sketch, 119, *119–120*; memory sketch, 113–114, *114*; prescriptive sketch, 119–121, *122–123*; referential sketch, 111–113, *112–113*; renderings, 116, *118*; sketch representation, 105–108; study sketch, 109–111, *111–112*; summary, 121–123

Smart Box concept, *53–54*

Solve 3D, 189

Spirula Speakers, 202

stages of new product development, 23–37; concept design, 24–30; concept development, 30–31, *31–33*; detail design, 31–37, *34–36*

STEP, 78

stereolithography apparatus (SLA), 189

STL, 78

storing sketches, 107, 111

storyboards, *27*

study sketches, *6–7, 22, 28, 30, 50–51, 57–58, 61–62, 86,* 109–111, *111–112, 209*

summary, 163–164, *163–164*

surface models, *78–79*

system prototypes, *172,* 178–179, *181, 200, 207–208*

T

"talk back", 4–6

talking sketches, 107

technical drawings, 137–140, *139–140*

technical illustrations, *91,* 140–141, *141–142, 211*

thinking sketches, 107, 110

third prototypes, 177

3D CAD, 6, xvii, 44, 50, *95–96, 195–198*

3D printing, 19, *20, 23,* 167

3D sketch models, *85, 148,* 150, *151*

3MF, 78

throwaway prototypes, 170–171

thumbnail sketches, 109

tooling prototypes, 182, 184, *184*

tools of design representations, 84–88

Turchi, Mario, 20

types of design representations, 81–84

U

USB port, *17*

V

vat photopolymerisation, 189

VERMEIL, 78

Viemeister, Tucker, 20

virtual prototyping, 167

visionary drawings, 127

W

Wallets, *8, 29–30, 33, 35–36, 62–63, 91, 120–121, 142*

For Product Safety Concerns and Information please contact our EU
representative GPSR@taylorandfrancis.com
Taylor & Francis Verlag GmbH, Kaufingerstraße 24, 80331 München, Germany

www.ingramcontent.com/pod-product-compliance
Lightning Source LLC
Chambersburg PA
CBHW061405210326
41598CB00035B/6107

*9 781032 131085 *